Theodore R. Schatzki
Martin Heidegger: Theorist of space

T0139772

SOZIALGEOGRAPHISCHE BIBLIOTHEK

Herausgegeben von Benno Werlen

Wissenschaftlicher Beirat:

Matthew Hannah / Peter Meusburger / Peter Weichhart

Band 6

THEODORE R. SCHATZKI

Martin Heidegger: Theorist of space

2nd unrevised edition

Franz Steiner Verlag

Umschlagbild mit freundlicher Genehmigung von François Fédier

Bibliografische Information der Deutschen Nationalbibliothek:
Die Deutsche Nationalbibliothek verzeichnet diese Publikation in der Deutschen
Nationalbibliografie; detaillierte bibliografische Daten sind im Internet über
<http://dnb.d-nb.de> abrufbar.

© Franz Steiner Verlag, Stuttgart 2017
Druck: Bosch Druck, Ergolding
Gedruckt auf säurefreiem, alterungsbeständigem Papier.
Printed in Germany.
ISBN 978-3-515-11761-6 (Print)
ISBN 978-3-515-11781-4 (E-Book)

For Louis and Helena, apples of my eye

Contents

Abbreviations

C	*Contributions to Philosophy (From Enowning)*
BDT	"Building Dwelling Thinking"
Ister	*Hölderlin's Hymn "The Ister"*
PHT	*History of the Concept of Time. Prolegomena*
SZ	*Being and Time*
OWA	"The Origin of the Work of Art"

Note on Quotations, Translations, and Cited Books

All quotations from Heidegger's works are taken from published English translations when these are available. I have freely amended these translations, in part in order to impose uniformity. The original German of these quotations is supplied in the appendix and can be located by way of the alphabetic superscript reference system. Translations of texts by Heidegger that have not been published in English are my own.

I use the titles of the English translations of Heidegger's works when they exist, appending the German title only when clarity requires it.

Because there are two quality translations of *Being and Time*, both of which list the German pagination, references to this book are abbreviated "SZ" and cite page numbers in the German text. Quotations from this book are based on the Macquarrie and Robinson translation.

All books and essays cited or mentioned in the text are contained in the bibliography in chapter six.

Preface

This book is about both Martin Heidegger as a theorist of space and the legacy of his ideas on spatial phenomena. Examining these topics is possible only in conjunction with a discussion of his philosophical ideas generally.

The book's two central chapters (chapters three and four) seek to describe and analyze Heidegger's ideas simply and straightforwardly. I hope that what I write is accessible and provocative to anyone at the university interested in its topics, from those who are either largely unfamiliar with his philosophy or outside the discipline of philosophy to scholars with professional interests in Heidegger, regardless of their field. It has become commonplace to describe a certain class of book as of interest to beginners and experts alike. In my case, I did not write with either the uninitiated or the professional—or any particular audience—in mind, though I was keenly aware that the book was supposed to be intelligible to multiple constituencies. As indicated, I just tried to state his ideas simply and straightforwardly. I did this in the belief, moreover, that when matters are put simply what is said should be understandable by all, informative to different degrees to those with varying amounts and types of prior acquaintance, and thought-provoking for those with well-developed opinions. What fuels my hope that proceeding in this way is not vain and delusional are my experiences teaching on the basis of this belief. I have found, for instance, that I can use the same descriptions of "what is basically going on" in a masterpiece such as Plato's *Republic* in introductory courses on moral and political philosophy and in graduate courses in social thought—though, of course, the contexts in which these descriptions are presented and the discussion they generate differ in the two cases. In any event, aiding the more inexperienced readers of this book are two brief initial chapters that outline Heidegger's life and his philosophical works. The final chapter on the legacies of Heidegger's thoughts in general and on spatial matters in particular should contain something for everyone.

This book was originally commissioned for a series of books on leading theorists of space. After three volumes had been written for the series, the publisher reneged on the deal, leaving the three books, of which this is one, in limbo. Two and half years passed before Benno Werlen, the editor of the would-be series, abandoned hopes of reviving it. At this point, Steiner Verlag, the publisher of Professor Werlen's series, *Sozialgeographische Bibliothek*, agreed to take on this volume as part of the latter series. The book's organization—the order, lengths, and overall subject matters of the chapters—adhere to the format that had been envisioned for the original series. When, after two and half years, the publication of the book finally became insured, there was little time, and I had little inclination, to alter the chapters' order, lengths, or topics. I did substantially revise the principal chapters, but the book as a whole still displays the format intended for the volumes of the defunct series.

I would like to thank Professor Werlen for standing by this book through the years and Steiner Verlag for agreeing to publish it. My graduate student, Brandon Absher, was a great help looking over the proofs. A large thanks is also owed to Hubert L. Dreyfus, from whom I initially learned Heidegger.

1 Biography

Martin Heidegger was born on September 26, 1889 in Messkirch, Germany, not far north of Lake Constance near the border with Switzerland. His family was composed of farmers and craftsmen. His father was a master cooper and sexton of the local Catholic church, whereas his mother was a farmer's daughter from a neighboring village. Heidegger received a classical education at a Jesuit Gymnasium in Greek, Latin, and German language and literatures. His philosophical education began in 1907 at age seventeen when the pastor of a church in Constance gave him a copy of Franz Brentano's *On the Several Senses of Being in Aristotle* (1862). This book made a strong impression on Heidegger, and his later philosophical work made the topic of being his own.

Heidegger entered the University of Freiburg in 1909, where he began studying theology under Carl Braig, having earlier consulted a book of Braig's titled *On Being: An Outline of Ontology* (1896). Later that year, Heidegger heard of a book that Brentano's student Edmund Husserl, the founder of the phenomenological movement, had written titled *Logical Investigations* (1900–01). Borrowing it from the library, Heidegger was so impressed by Husserl's phenomenology that he kept the book in his room for two years, "read[ing] it again and again." In 1911, he dropped theology and elected philosophy as his main area of study. Even at this point his mentors recognized his exceptional abilities. In place of a church scholarship he had forfeited by virtue of leaving theology, he was provided a small grant.

Heidegger's first published philosophy article appeared in 1912 under the title "The Problem of Reality in Modern Philosophy." He received his Ph.D. in 1913 with a dissertation titled *The Theory of Judgment in Psychologism*. His dissertation advisor was the leading neo-Kantian German philosopher Heinrich Rickert. Between 1910 and 1914, Heidegger immersed himself in the study of various thinkers and poets, including Friedrich Hölderlin, Friedrich Nietzsche, G.W.F. Hegel, Wilhelm Dilthey, Søren Kierkegaard, Georg Trakl, Rainer Maria Rilke, and Fyodor Dostoevsky. Ideas of Hölderlin, Nietzsche, Dilthey, and Kierkegaard were appropriated in his subsequent thought. When war broke out in 1914, Heidegger volunteered for military service, but his weak heart quickly led to his being assigned to the military postal service in Freiburg. This turn of events enabled him to continue his studies. By 1915–16 he had completed his postdoctoral dissertation, the *Habilitation*, titled *The Theory of Categories and Meaning in Duns Scotus*.

In 1916 Rickert accepted an appointment as successor to Wilhelm Windelband at Heidelberg. Rickert's own successor was Husserl. In 1917, Heidegger reentered military service. He was initially stationed once again in Freiburg. This enabled him to continue his academic work and to begin his teaching career. Later he was sent to a meteorological station on the Western front near Verdun, where he served until the end of the war. During this period he married Elfride Petri, the daughter of an imperialistic and conservative Junker officer. The couple had two sons (1919, 1920).

After the war, Heidegger's own ideas and academic career gathered steam. He became Husserl's Assistant in 1919. Although he lectured under the title "Phenomenology" from 1917 to 1926, he began to articulate his own, powerful voice. The influence of Dilthey's studies of historical life combined with those of Husserl's transcendental phenomenology, Greek inquiries into ontology and logic, and Christian thought to yield the seeds of an original approach to questions of being.

In 1922, at the relatively young age of thirty three, Heidegger became an *Extraordinarius* professor at Marburg University. He lectured there 1923–28. He held forth primarily on the history of philosophy, though he also gave lectures on time, logic, and truth. He spoke wearing a loden jacket and knickerbockers. Even though he had published nothing since the middle of the previous decade, his reputation, based on teaching alone, grew in leaps and bounds. Hannah Arendt, a student of Heidegger's with whom he had a passionate affair from 1924 to 1933 (despite tremendous risk to his career) and who went on to become one of the very most distinguished political think-ers of the 20th-century, described his reputation as follows:

> Thinking has come to life again; the cultural treasures of the past, believed to be dead, are being made to speak, in the course of which it turns out that they propose things altogether different from the familiar, worn-out trivialities they had been presumed to say. There exists a teacher; one can perhaps learn to think.[1]

In the winter 1925/26 semester, the philosophy faculty at Marburg nominated Heidegger to be Nicolai Hartmann's successor as the prin-cipal chair of philosophy. The ministry in Berlin rejected the ap-pointment because Heidegger had not published a book since his *Habilitation*. Luckily, Heidegger had an unfinished manuscript called *Being and Time*. Husserl arranged for its publication. Page proofs sent to the ministry were returned, however, with the remark "Inade-quate." The ministry relented in 1927 when the book was published. Despite its hasty publication and incompleteness, it quickly became recognized as a philosophical work of the highest order.

In 1928, Heidegger succeeded the retiring Husserl in Freiburg. From then until 1933 he lectured on themes arising out of *Being and Time* and on the history of philosophy. In 1929, there occurred the famous Davos Disputation between Heidegger and Ernst Cassirer, in which Heidegger's ontological ideas, grounded in an analysis of hu-man existence, collided with Cassirer's neo-Kantian and atemporal epistemological theories. Twice in the early 1930s Heidegger declined a professorship in Berlin. In April of 1933, after the Nazis had come to power, he accepted the position of rector at Freiburg, to which the combined faculties had elected him. The motivations for this move are controversial—opportunism; rural Catholic conservative upbring-ing and conservative convictions; weariness in the face of political, economic, and spiritual decline in Germany combined with faith that the Nationalist Socialist movement could renew the nation; the desire to be the intellectual *Führer* to the *Führer*; these are some of the most commonly cited reasons. In May of that year, he joined the Nazi party. His famous inaugural speech as rector (the *Rektoratsrede*) called on the German university to assume new, philosophically con-

---

[1]   Hannah Arendt, "Martin Heidegger at Eighty," *New York Review of Books,* October 21, 1971, p. 51.

ceived epistemological duties as part of a general renewal of German society. Heidegger abandoned the rectorship ten months after he had accepted it (in part because he refused to fire two anti-Nazi professors), but the damage to his subsequent reputation had been done. Stories about nasty behavior later circulated; after the war he claimed to have abandoned the Nazi party in 1934 although it turned out that he had remained at least a sympathizer if not also a member into the 1940s; after the war he never addressed or directly acknowledged, let alone apologized for, the ghastly crimes of the Nazis. Despite the fact that the Nazis criticized Heidegger in the 1930s and placed restrictions on his lecturing, publishing, and travel, such matters secured him lasting infamy, guaranteed continuing controversy (the most intense scholarly engagement with his Nazi involvement occurred in the 1980–90s), and led some scholars to refuse to address his ideas.

After he resigned the rectorship, Heidegger continued teaching about logic, truth, and metaphysics as well as the history of philosophy. These topics were supplemented, however, by new ones connected with Hölderlin and Nietzsche. Heidegger had become acquainted with the writings of this poet and this thinker while still a student, but not until the mid 1930s did they seem appropriate material for teaching. Heidegger continued lecturing on their work and the themes they raised well into the Second World War.

After the war, the French occupation banned him from university teaching after a denazification hearing (also requisitioning his house and library). The ban was lifted in 1951, one year before his scheduled retirement. Between 1946 and 1951 he gave private seminars and published his first philosophical essays since around 1930. Heidegger spent the 1950s and 1960s writing, thinking, occasionally offering seminars or public lectures, and overseeing the publication of a long line of prominent works. Heidegger's final official lecture course in 1951 was titled *What is Called Thinking?* He returned to the university as emeritus professor in the mid 1950s to deliver two lecture series, one titled *The Principle of Reason* and the other on Hegel.

During the 50s and 60s, Heidegger traveled to France and Greece, but mostly "stayed in the provinces," in southwestern Germany, Freiburg and Messkirch. Many hours were spent at his ski cabin in the Black Forest at Todtnauberg, where he often received visitors, many of them young philosophers. He also cultured "friendships" with a number of leading intellectuals, including Hannah Arendt, the physicist Werner Heisenberg, the theologian Rudolf Bultmann, the poet René Char, and the painter George Braque, though many of these friendships were troubled and haunted by the war and Heidegger's Nazi involvement. His enigmatic meeting with the German-Jewish poet Paul Celan in 1968 is symptomatic of these troubles. Heidegger's final public performance was a seminar in 1973 at Zähringen. In 1972 he made plans for the publication of all his writings. In May 1976 he died at the age of eighty six and was buried in Messkirch, between his mother and father, in a cemetery he had passed daily as a boy.

2   Philosophical Works

Heidegger's work spans six decades, from his earliest student articles and reviews in the 1910s to the final seminars and lectures in the 1970s. His production is prodigious. In addition to the many works published in his lifetime, his corpus includes a considerable number of talks, lectures courses, and unpublished texts (some book-length), most of exceptional care and quality. His projected Collected Works encompasses 102 volumes, a large number even by German standards.

Heidegger's first lecture was delivered in 1910 about Abraham a Santa Clara (1644–1709), a court preacher who had been adopted as a role model by contemporary conservative elements of the Catholic church. This lecture evinces Heidegger's affinity for Catholic rural conservative antimodernism, an affinity he never abandoned. Indeed, Heidegger's work never shook off the philosophical and cultural sense of crisis accompanying this antimodernism. In 1911–12, he continued publishing articles, reviews, and poems in *Der Akademiker*, the journal of the German Association of Catholic Graduates. His first philosophy essay was "The Problem of Reality in Modern Philosophy" (1912). This essay criticizes the preoccupation of modern philosophy with epistemology, holding that philosophy should become relevant to scientific culture by formulating new problems and stimulating knowledge in the natural and historical sciences. His doctoral dissertation of 1913, *The Theory of Judgment in Psychologism*, argued that although psychology is unable to analyze judgment, which is a phenomenon of logic, logic is nonetheless dependent on extralogical contexts, namely, everyday experience, history, and metaphysics. Similar themes mark his postdoctoral dissertation, *The Theory of Categories and Meaning in Duns Scotus* (1915–16). This work combines the study of pure logic with an appreciation of medieval history and culture, arguing that the proper content of all logical problems is historical. This same move away from logic toward history is evident in his inaugural lecture of July 1915, "The Concept of Time in the Science of History," which distinguishes the time of modern physics from the time of history.

Returning to academic work after World War One, Heidegger concentrated on his lectures series, offered occasional public lectures, and worked on manuscripts that eventually became his magnum opus. From 1919 to 1923, he lectured at Freiburg as Husserl's Assistant, labeling most of his lecture series phenomenological, for example, *Phenomenology and Transcendental Value Philosophy* (1919), *Phenomenology of Intuition and Expression* (1920), and *Phenomenological Interpretations of Aristotle* (1921/22). These lectures are a baroque amalgam of Husserl's phenomenology, Dilthey's hermeneutics of historical life, Scheler's philosophical anthropology, and Christian ideas of St. Paul, Augustine, and Martin Luther. Heidegger described human life as a finite, situated, and factical (*faktisch*) flow and advocated phenomenology as the science of the nonobjectifiable grounds of life and experience. This analysis received more systematic—though anything but definitive—form in his last Freiburg lectures, *Ontology— The Hermeneutics of Facticity* (1923). Several key themes of *Being and Time* were enunciated in this period, including factical existence, authenticity, the destruction of metaphysics, and *Dasein* (this word at once denotes human beings and expresses their being).

In 1923, Heidegger moved to Marburg. No longer Husserl's assistant, he stopped titling his lectures phenomenology, though two of his later Marburg lecture series were so labeled (*The Basic Problems of Phenomenology* [1927] and *Phenomenological Interpretation of Kant's Critique of Pure Reason* [1927/28]). It seems that Heidegger, long sensitive to the historical character of human life, had recognized the historical contingency of Husserl's allegedly atemporal transcendental investigations. Heidegger's approach now became hermeneutical: an articulation, or working out, of human being from within its midst. His lectures ranged over an impressive range of topics: the beginnings of modern philosophy, Aristotle's *Rhetoric*, Plato's *Sophist*, the nature and history of time, logic and truth, intentionality, the basic concepts of ancient philosophy, the history of philosophy from Aquinas to Kant, Kant, and the metaphysical grounds of logic. The question of the meaning of being was first formulated in stable form in his 1924

Aristotle lectures. Sometime toward the beginning of his Marburg period (or earlier), Heidegger had also begun writing a book manuscript. This manuscript, of which there are two preliminary runthroughs, was published in 1927 as the epochal *Being and Time*. It was his first publication since 1917.

Heidegger returned to Freiburg in 1928. A second book, *Kant and the Problem of Metaphysics*, was published in 1929. This book somewhat carries out the first stage of the destruction of philosophical history that had been announced, but never carried out, in his magnum opus. By this point, moreover, Heidegger's originality and intellectual daring were in full display. His public lectures of the next few years were widely read both at the time and later, for example, "What is Metaphysics?" (his inaugural lecture as professor of philosophy at Freiburg in 1929), *The Essence of Reasons* (1929), and "On the Essence of Truth" (1943, delivered publicly in 1930). These texts largely extend the framework of *Being and Time*, focusing more squarely on nothingness, grounds, freedom, and truth. His lecture courses at this time either extended the ideas of *Being and Time*, examined Greek philosophy, or took up postKantian German idealism, a nexus of thought on which he had not lectured since 1916.

At the same time, a fundamental shift in his thought was underway. Considerable controversy exists among Heidegger scholars on whether a significant reorientation of Heidegger's thought occurred sometime between the publication of *Being and Time* and 1934. Some scholars date this reorientation to his 1930–31 lectures on truth and Plato. Others deny that a major shift occurred. In my opinion, fundamental changes have clearly emerged no later than Heidegger's first lecture series on Hölderlin in 1934. Most prominent among these is the shift from the meaning to the truth of being (see the following chapter).

The 1934 Hölderlin lecture series also introduces or reinforces new themes: poetry, language, place, gods and humans, strife, and the relation of these to time and historicity (*Geschichtlichkeit*). Although Heidegger continued lecturing on familiar topics and figures (e.g., metaphysics, Kant, and Schelling), in 1936 he also began to discuss Nietzsche; from 1936 to 1941, six of his eight lecture courses were on Nietzsche (though the last, in 1941/42, was replaced at the last moment by a series on thought and poetry). Considering their historical context, the themes of these courses were timely and monumental:

power, will, nihilism, art, strife, metaphysics, and being. Two further
courses on Hölderlin appeared in 1941–42 after his encounter with
Nietzsche. Heidegger's intense engagement with Hölderlin and
Nietzsche reflected the convictions that both writers hold the key to
comprehending the contemporary world and that Hölderlin also provides concepts for grasping the task of getting beyond it.

The period of Nazi domination gave rise to several important, and
only later published, essays: "The Origin of the Work of Art" (1935/36;
see next chapter), "The Age of the World Picture" (1938), "Plato's
Theory of Truth" (1940), and "On the Essence and Concept of Φνσις
[phusis], Aristotle *Physics* B1" (1939). In 1936–38, Heidegger also
composed what today some scholars describe as his second magnum
opus, *Contributions to Philosophy (From Enowning)*. This book—the
only systematic exposition of his thought after *Being and Time*—
focuses on the concept of the event (*Ereignis*-translated as "enowning") as the new lynchpin in Heidegger's account of being, humanity,
Dasein, language, thinking, and the modern world. The book is dire
in its depiction of the abandonment of being in the contemporary
world.

After the war Heidegger gave only a handful of lecture series, *What
is Called Thinking?* (1951/52) being the most famous and last official
one. A remarkable line of publications began with the "Letter on Humanism" (1947), which is an excellent introduction to his later
thought. This essay responds to questions that Jean Beaufret, a French
intellectual with whom Heidegger became friendly during the occupation, had put to him concerning the famous essay of the French exis-
tentialist Jean-Paul Sartre, "Existentialism is a Humanism." Beaufret
had a major hand in fomenting the considerable French interest in
Heidegger's work that blossomed after the war.

In the 1950s, the Nietzschean strife so evident in Heidegger's 1930s
work was subsiding (though Heidegger's famous essay on the character and danger of the age of modern technology, "The Question Concerning Technology," appeared in 1953–54). His Hölderlinian side
took charge. Dwelling (*Wohnen*) and releasement, or letting be (*Gelassenheit*), joined thinking and poeticizing (*Dichten*) as part of a redemptive vision for Western humanity (alongside Heidegger's continuing inquiries into logic, judgment, and being). The publications
from this era include *Off the Beaten Track* [*Holzwege*] (1950), "Building Dwelling Thinking" (1952), *Vorträge und Aufsätze* (1954), *The*

Question of Being (1955), *The Principle of Reason* [*Der Satz vom Grund*] (1957), *Identity and Difference* (1957), *Discourse on Thinking* [*Gelassenheit*] (1959), and *On the Way to Language* (1959).

Heidegger's intellectual activities began dying down in the 1960s. He delivered fewer lectures, and fewer publications appeared. His 1930s Nietzsche lectures were published in 1961. His important late essay "Time and Being"—which is profitably read alongside *Being and Time*—appeared in 1969 in a collection titled *On Time and Being* (*Zur Sache des Denkens)*]. He also held several seminars, including a well-known one with Eugen Fink in 1966/67 on Heraclitus. In 1966, he conducted an interview with the news magazine *Der Spiegel*, which, at his request, was posthumously published, in May 1976.

3 Basic Philosophical Ideas

The decisive issue in Heidegger's thought is the question of being. "The" question of being is actually several questions. To speak of "the" question of being is to indicate that Heidegger's philosophy is overwhelmingly preoccupied with being.

For present purposes, I shall divide Heidegger's thought into three phases: one extending into the early 1920s, a second—the *Being and Time* phase—extending from the early 1920s into the early 1930s, and a third covering the early 1930s to the 1970s. This periodization does violence to Heidegger's thought from the early 1930s onwards. As sketched in the previous chapter, his ideas in the 1930s diverge, sometimes strongly, from his thoughts in the 1950s. For the purposes of the present book on space, however, the damage is minimal.

The Questions of Being

What is being? The question is not easy to answer straight out because Heidegger's ideas on the matter changed. Moreover, we use the forms and cognates of the verb "to be" so often and so unreflectively that the topic of being might seem trivial and obvious. At the same time, the fact that being is so pervasive, so "near," also entails that the topic is likely to seem elusive when attention is called to it. What *is* being? In the first two phases of his thought, Heidegger would have replied that being is what is said whenever it is said that something is: that it is, full stop (existence), that it is this (identity, x=y, e.g., "John is the person who won the lottery"), and that it is characterized by such and such (predication, e.g., "the sky is blue").

Heidegger claimed that the question of being has been forgotten. In the first two phases of his thought, he further claimed that one sign of its oblivion is the fact that, throughout history, Western philosophers have said such things as that being is the most universal concept, that it is the emptiest concept, and that it is self-evident. In effect, these claims are ways of putting the topic of being aside while tackling more tractable matters such as knowledge, truth, justice, and the good. Heidegger's philosophy aims to reawaken the question of being. According to Heidegger, moreover, not just the incisiveness and future of Western philosophy are at stake in such a reawakening. The condition and fate of the Western world are also at issue. Heidegger believed this because he thought that the working out of certain metaphysical ideas has underlain the course of Western history, that the completion of this working-out process is responsible for the ever growing systematization characteristic of the contemporary world, and that humanity can escape this development only by re-engaging the issue that these metaphysical ideas covered up—the question of being.

Over the course of Heidegger's career, the specific questions that were "the" question of being changed twice. While a student and Husserl's assistant at Freiburg (1909–1922), the question of being consisted of two traditional metaphysical issues: What is a being (*Seiendes*)? and What is the unity of the many ways in which beings can be understood or said in their being? These questions are two of the leading "ontological" issues bequeathed to the philosophical tradition by Plato and Aristotle, who are regularly considered to be—and Heidegger concurred in this judgment—the founders of systematic Western philosophy. (An "ontological" question is one pertaining to being.) The first question asks what it is for an entity to be. The second question presupposes Aristotle's observation that being can be said in many ways: substance, quantity, quality, place, time, etc. The question arises as to whether this multiplicity is unified. Later in his career, Heidegger claimed that to pose these questions is to treat being metaphysically, hence, to be subject to his later criticisms of metaphysics.

In this first, "conventional" phase, Heidegger's thoughts on his leading question had a strongly universalist and apodictic character; they were exercises in pure logic that aimed for absolute truth and certainty. At the same time, Heidegger evinced considerable sensitivity to the historical and cultural context of ideas about being and logic.

This sensitivity probably arose from (1) the historicism that was pervasive at that time (historicism is the doctrine that what something is depends on its historical context) and (2) his reading of Hölderlin, Nietzsche, Kierkegaard, and Dilthey. In his postdoctoral dissertation of 1916, for instance, Heidegger argued both that a philosophical account of knowledge (judgment) must be based on a pure logic of concepts and that history is the context for all logical issues.

In the second phase of Heidegger's career, the centerpiece of which is *Being and Time*, "the" question of being became, What is the meaning of being? This question is of Heidegger's invention although, as he himself later recognized, posing it presupposes certain metaphysical ideas. In this second phase, the mix of the absolute and the historical shifts. After World War One, Heidegger became critical of the metaphysical tradition that began with Plato. He turned away from Husserl's ahistorical treatment of consciousness and the transcendental subject as a path toward answering the question of being and began analyzing ongoing, matter-of-fact, historical life. He did this in the belief that what he called "factical" (*faktisch*) life underlies metaphysics and is what must be examined in order to grasp the meaning of being. Yet, even though Heidegger turned to historical life, his analysis of it exhibited strong absolutist features. *Being and Time* is resolutely abstract and the scope of its claims are universal.

In the third phase of Heidegger's reflections on being, which began around 1929 but the fuller contours of which first began to become clear around 1934, the question of being became, What is the truth of being? Once again, this is a question of Heidegger's invention for which there is no precedent in Western philosophy. As we shall see, the transition from the *meaning* to the *truth* of being was of great consequence. In this third phase, moreover, the historical dimension of Heidegger's endeavor became dominant. Most of his lectures and essays interpreted past philosophers and poets as part of a broader attempt to grasp the history of the truth of being. Systematic abstract conceptualization, although still important, became less prominent.

As indicated, the question of being was Heidegger's leading issue. Above all in his lectures and essays from the 1930's onwards, a certain predictability affixed his discussions: whatever topics and figures are addressed in these lectures and essays are invariably related to the question of being. All manner of doctrines of past philosophers were scrutinized for implications and presuppositions about being, the

poems of leading Germanic poets were transformed into meditations
either on or bearing upon being, the contemporary world was understood as a moment in the history of being, and all Heidegger's con-
structive thinking concerned being. Heidegger often remarked that
every great philosopher had one thought; his own work strikingly
illustrates this claim. Indeed, the reader can get the impression that
philosophy is ultimately and only about being. But although whatever
Heidegger addressed he did so for the sake of the question of being,
being is not all he discussed. Indeed, the range of topics he examined
is unusually broad, including the nature of truth and reality, the character of logic, the nature of human being, knowledge, as well as coexistence, death and anxiety, time and space, fate and destiny, the history
of philosophy, the nature of history, divinity, technology, language,
the contemporary world, the fate of humanity, the subject-object relation, nihilism, and humanism. The only prominent areas of philosophy he did not tread are ethics/morality and political philosophy,
though, as we shall see, he did discuss and advocate a particular ethics
(understood as a way to live, as opposed to morality construed as a set
of precepts to obey). That Heidegger was able to relate so many matters to the question of being, and thereby tie them together into a unified (if evolving) conceptual package, is a tribute either to his profun-
dity or to his monomaniacalness, depending on your point of view.

Being and Time: Existence, World, and Temporality

I pass over phase one of Heidegger's engagement with being. One
reason for this is that at this early point he had not yet found his own
voice. Another reason is that it is the ideas of the second and third
phases that have inspired and exercised subsequent philosophers.

     As noted, in phase two of his career the question of being took the
form, What is the meaning of being? Heidegger used the expression
"meaning" (*Sinn*) technically: the meaning of something is that
wherein its intelligibility (*Verständlichkeit*) maintains itself (SZ 151).
It so happens, moreover, that something, for example, a rock, a person, or a poem, is not intelligible on its own: its intelligibility is not a
property it possesses independently of everything else. Rather, something is intelligible *to* human beings. Only to a person is a rock intelligible, say, as a paperweight (or as a rock, for that matter). Likewise,

only to a person is another person intelligible as a friend and writing intelligible as an ode to an urn. That intelligibility is relational is suggested by the fact that another word for intelligible is "understandable:" for X to be understandable as Y is for X to be understandable as Y *for p* (however unspecified p might be).

The relationality of intelligibility entails that, when Heidegger asked about the meaning of being, he in effect posed a question about the understanding of being. He asked, not about being flat out, but about the intelligibility, thereby, the understanding of it. Now, it so happens that Heidegger claimed that it is distinctive of the sort of creatures "we" are to possess an understanding of being (where "we" presumably refers to anything capable in principle of reading his work, thus, roughly, people). In asking about the understanding of being, Heidegger thus asked about something definitive of people. It follows that the question of being is a question about something in so far as it is understood by "us." Humanity and its existence thereby become part of what is thought when thought ponders being.

Indeed, because "we" are distinctive in having an understanding of being, Heidegger argued that the first step in uncovering the meaning of being is to analyze "our" being. Heidegger referred to "us" with the word *Dasein*. (Because there is no equivalent for this in English, in the following I will use the German word.) Accordingly, the first task is to analyze the being of Dasein. The second task is to use this analysis to show that the meaning of being, that within which its intelligibility maintains itself, is time. Time is the "horizon" within which being as being is understood: however being is said or grasped, it is understood in terms of time. The third task, finally, is to go back through three key stages of Western philosophy (Kant, Descartes, and Aristotle), using the thesis that time is the meaning of being (1) to dismantle the tradition that has covered up the question of being and been conceiving being in terms of time even though it did not understand this and (2) to arrive back at the sources that gave rise to that tradition.

It turned out, however, that *Being and Time*, the book that was projected to execute all three tasks, carried out the first alone. The second task was never carried out, and the third was at best partially executed in a subsequent publication on Kant (although Heidegger occasionally lectured on Kant and Aristotle in phases two and three of his career). Hence, Heidegger's magnum opus, the book that is as philosophically influential as any single text of the 20th-century, turned out not to be a book about the issue that dominated everything he wrote and taught, but a book primarily about human existence. Indeed, it is precisely due to its account of human existence that it has become so famous.

Heidegger argued that Dasein is the only entity whose essence (*Wesen*) does not consist in the possession of specific properties. Something is, say, a rock or a pencil if it possesses those specific features the possession of which qualify something as a rock or pencil. By contrast, "the essence of Dasein lies in its existence."[a] (SZ 42) This means that the essence of Dasein lies in its to-be (*Zu-sein*), which in turn means that its essence lies in having a self-understanding that is worked out by living it. (This is a version of Kierkegaard's idea that the subject is a relation to itself.) In other words, *who* a given person is is fundamentally open and filled in by that person's leading a particular life. Whereas, consequently, *what* something other than Dasein, for example, a rock or pencil, is does not vary among rocks or pencils, *who* a person is varies among persons according to their self-understandings and how these self-understandings are worked out through existing. Even regarding people all of whom understand themselves the same, say, as chefs, who each is varies according to the different way each lives out that understanding of him- or herself that is linguistically articulated with the same word.

A guiding intuition in modern philosophy since Descartes had been that people (Heidegger's "we") are subjects, something "in here" opposed to the world "out there." These subjects were regularly conceptualized as self-contained inner entities that can in theory exist independently of the world. Because of this conception, modern philosophy has often been labeled "Cartesian" and said to be based on a subject-object split. It is easy to see how such issues as the existence of the external world and the possibility of knowing it came to exercise the ingenuity of the profoundest modern thinkers (such as Descartes, Hume, and Kant).

Heidegger overturned this tradition by claiming that the entity "we" are is not encapsulated in an inner sphere standing over against the outer world. Rather, the entity that each of us is is, essentially, *in-the-world*. The central constitutive feature of human existence, of the being of the entity whose self-understanding is worked out by living, is *being-in-the-world*. It is difficult to convey the flash of insight that accompanied this claim as it spread through the continental European intellectual world. Along with the ideas of, above all, the German thinker and essayist Nietzsche and the Austrian-English philosopher Ludwig Wittgenstein, this notion sounded the death knoll for the "Cartesian" conception of the human being. Today the demise of this conception is pervasively affirmed in Western philosophy (at least in the letter, if not in the substance).

Heidegger had more surprises up his sleeve. In the course of describing the different ways people are in-the-world, Heidegger argued that the practical way of being-in, in which people are busy using equipment (*Zeug*, e.g., a telephone, a steering-wheel, or—Heidegger's favorite example—a hammer), has priority over the cognitive way of being-in, in which people acquire knowledge of things (and can theorize about them). Knowing (*Erkennen*), as he put it, is founded in the practical mode of being-in that consists in using things while carrying out particular projects for the sake of certain ends. The practical mode of comportment has priority over the cognitive mode in two ways: (1) in that the "default" mode of being-in is practical engagement, meaning that this is the mode a person is usually automatically in, and (2) in that a person cannot be capable of cognizing things without being capable of using equipment. In *Being and Time*, Heidegger also contended that a person comes to acquire knowledge only when a breakdown occurs in practical engagement and she must look at and scrutinize things; he seems, however, quickly to have abandoned this claim. This thesis of the priority of the practical was of great significance for 20th-century thought, and the issues it raises remain the focus of considerable discussion today. The thesis was nothing short of revolutionary because modern philosophy, presupposing the subject-object split, had made epistemology (theory of knowledge) the central philosophical endeavor. As suggested, questions of whether and how people can know the world and each other had been paramount from Descartes right up to Heidegger's mentors, Rickert and Husserl. The claim that knowledge is a founded mode of being-in shifts philosophical attention from knowledge per se to the practical forms of activity against the background of which knowledge exists and is acquired. In arguing thus, Heidegger joined ranks with such contemporaries as Wittgenstein and the American pragmatist John Dewey, who likewise challenged the centrality of epistemology in the name of practical activity.

This reorientation also led Heidegger to reconceive the world in which Dasein is (the world of being-in-the-world). Cartesian metaphysics conceives of the world as a collection of spatially extended objects (over against the subject). This conception of world underlay Newtonian physics and the development of modern chemistry, biology, and astronomy. According to Heidegger, by contrast, the world in which a person firstly and mostly (*zunächst und zumeist*) exists is a world of organized equipment that are available for his actions and projects: in ongoing life, the world a person inhabits encompasses a nexus of available equipment that help delimit the actions and paths he can take (the world also embraces other people, who are in-the-same-world that he is in). The world in which people dwell is not a collection of spatially extended objects. Of course, most of the things with which people have to do (e.g., telephones, hammers) are also spatially extended objects. But firstly and mostly—barring breakdowns in ongoing activity—people have to do with them, not *as* extended objects, but *as* equipment. Indeed, to encounter them as extended objects takes cognitive and experiential readjustment. This reconception of the world has immense consequences for the space of human life, as will be discussed in the following chapter.

Another signature Heideggerian thesis centers on the neologism, *das Man* (the One, the anyone, the they). In daily life, Heidegger claimed, most people are like anyone: they do what one (*man*) does, they say what one says, and so on. They are conventional, in other words, hewing to the norms of thought and action that are upheld at their time and place. A life unreflectively observant of social norms is an inauthentic (*uneigentlich*) existence. When living thus, people are not themselves, that is, they are not something of their own (choosing). Society, not the individuals involved, is responsible for what they do and think. In contrast, when a person explicitly takes responsibility for who she is and, cognizant of how things (including norms) stand, chooses a path in life, she pursues an authentic (*eigentlich*) existence. Living authentically does not require striking out on an unconventional path; a person can authentically observe the going norms. What makes existence authentic is that a person, cognizant, among other things, that she is capable of choosing herself and that ultimately no one but she can choose for herself, takes responsibility for her life, thereafter either proceeding unconventionally or resuming a life of observing norms.

Heidegger's account of authentic existence is largely presented in the second half of *Being and Time*. Although this account has not enjoyed nearly the sort of positive reception in the intellectual community that his analysis of being-in-the-word has, it has exerted considerable influence and continues to instigate scholarly debate. Its resonance derives partly from the role Heidegger ascribed to anxiety (*Angst*) and death in authentic existence. People who live inauthentically conduct their affairs unaware that they are inauthentic and that there is an alternative. Only in bouts of anxiety do they have an inkling of what is going on. In anxiety, the world becomes meaningless; a person becomes anxious about the mere fact that she exists, in the face of the mere fact that she does. Those experiencing anxiety thereby gain access to the realization that nothing, really, grounds their existence. The flip side of this insight is that a person's existence is the one thing that is truly hers, for which she can take responsibility. The realization that existence is one's own is brought home in confronting death. Inauthentic people flee death, treating it as an event that will undoubtedly befall one but that primarily concerns others. An authentic person faces up to the fact that death, as *her* possibility of having no more possibilities, is *her* own and, moreover, constant possibility. It is her ownmost (*eigentlichst*) possibility. The realization that there is something that is constantly and necessarily one's own and no one else's awakens a person to the possibility of a life of her own, a life where she takes responsibility and chooses herself. The person who does this is resolute (*entschlossen*): the authentic life is one of resoluteness (*Entschlossenheit*).

Talk of authenticity, anxiety, and death puts Heidegger into the company of such existentialists as Kierkegaard before him and Jean-Paul Sartre and Albert Camus after him. It has led phase two of his philosophy to be labeled "existential" and has, in no small part, contributed to Heidegger's popularity among students. At the same time, these are topics that Heidegger hardly discussed other than in *Being and Time*. What much more dramatically shifted the philosophical landscape was his analysis of being-in-the-world .

Few of the above topics directly bear on the question of the meaning of being, the ostensible topic of the projected book of which the extant *Being and Time* is part one. Heidegger did, however, work toward this question in this text. In the fifth chapter of division one, Heidegger analyzed the being-in dimension of being-in-the-world. He argued that being-in has three components: attunement (*Befindlich-keit*), falling (*Verfallen*), and understanding (*Verstehen*). Attunement is how someone finds herself already to be, whereas falling is what she is up to and involved with, and understanding is the projection of possible ways of being. (It is in his discussion of understanding, incidentally, that Heidegger developed the famous idea that all understanding and interpretation occur within a fore-structure that conditions them. This idea has disseminated throughout the world of arts and letters, the most famous appropriation being Hans-Georg Gadamer's argument that all understanding and interpretation are bound to history and tradition.) Ongoing life always embraces how things already are, doings and concerns, and possible ways of proceeding.

In chapter four of division two of *Being and Time*, Heidegger reinterpreted the three components of being-in as dimensions of time, or more exactly, as the three dimensions—past, present, and future—of the temporality (*Zeitlichkeit*) of human existence. Human existence is, ultimately, a temporal phenomenon: the meaning of Dasein's being, that wherein it maintains its intelligibility, is temporality. In chapter five of division two, moreover, Heidegger focused on the past and how human existence is intrinsically historical. It is in this (and the previous) chapter that he made the important point that the three dimensions of temporality (but not the past, present, and future of time) are simultaneous. The existential past is not something that is no more, just as the existential future is not something that is not yet. Rather, just like the existential past, the existential past and future, what is already the case and what is possible, *are* so long as Dasein is; they do not trail off behind Dasein or await it ahead. *Being and Time* then concludes with an argument that world time, the time of things and events in the world, as well as clock time, understood as a flow of nows, are derivative from the temporality of existence. This thesis parallels the theses of the priority of the practical over the cognitive and of the priority of the world as equipment nexus over the world as collection of extended objects.

Being and Time, however, never proceeds beyond an examination of Dasein's being. The transition from temporality as the meaning of Dasein's being to time as the meaning of being never occurs. (See, however, the brief discussion at the conclusion of *The Basic Problems of Phenomenology*, a lecture series delivered shortly after the publication of *Being and Time*.) It is as if Heidegger realized that the link that he had forged between a kind of time, to wit, temporality, and the meaning of a kind of being, namely, Dasein's, led to a dead end: it could not be generalized to the meaning of being period. Somehow, in other words, Heidegger had taken a wrong turn when he chose to focus on Dasein. To suggest what misfired, we need to introduce a fateful concept that has hitherto been passed over.

In chapter five of division one, Heidegger introduced the notions of the there (*Da*) and the clearing (*Lichtung*). The notions are, for our purposes, interchangeable. The there-clearing is a space in which things are, in which things show up for people. It is something like an open (*Offene*), presupposed by all determinateness and representation, where anything that is is. Although Heidegger's discussions of them are brief, the concepts are clearly of the greatest importance—the clearing is the clearing of being. Moreover, Heidegger averred that "the being that is essentially constituted by being-in-the-world *is* itself its there."[b] (SZ 132). The word *Dasein* does not just refer to people, but also expresses their being: being there, being the there. Dasein, in other words, is the clearing. More precisely, the clearing is cleared by the three components of being-in, attunement, falling, and understanding. Understanding is primary in this regard: Dasein's understanding clears (*lichtet*), or illuminates (*erleuchtet*), the clearing of being. Indeed, the natural light (*lumen naturale*) of Dasein's understanding *is* the clearing. In division two of *Being and Time*, Heidegger transformed this claim into the thesis that, ultimately, the temporalizing (*Zeitigung*) of temporality is what clears the clearing.

Historically speaking, such claims align Heidegger with Kant as a kind of idealist philosopher. According to Kant, possible objects of experience instantiate certain basic conceptual determinations called "the categories of the understanding," which are innate in the mind. Any object of experience substantializes a number of these categories. Similarly, Heidegger in effect argued that anything that is instantiates Dasein's understanding, in particular, Dasein's understanding of being: anything that is possesses a being of a sort Dasein understands— else it is not. Just as in Kant, consequently, entities—their natures, not their existence—depend on us. Without categories of the understanding, no objects of experience; without understanding of being, no entities. Kant called his position transcendental idealism. Heidegger preferred such phrases as "phenomenological ontology."

After *Being and Time*: Truth, Metaphysics, and the Event

As noted, Heidegger could not find an argumentative path from temporality as the meaning of Dasein's being to time as the meaning of being. He concluded that the problem lay (at least partly) in the idealist character of his own enterprise, in the presumption that the clearing is cleared by, thus at once one with, Dasein's understanding (or temporality). As discussed, however, construing the question of being as a question of the *meaning* of being makes the question of being eo ipso a matter of Dasein's understanding. So long, consequently, as Heidegger interpreted the question of being thus, he could not set aside the equation of the clearing with Dasein's understanding (or with temporality). Heidegger's leading question henceforth changed from the meaning of being to the truth of being, where truth was understood as unconcealedness (*Unverborgenheit*). (*Unverborgenheit* is Heidegger's controversial interpretation of the Greek word *alḗtheia*, which in English literally means unforgetfulness or unconcealing but is regularly translated as "truth"). The clearing that Heidegger fleetingly discussed in *Being and Time* subsequently came onto center stage and became the open of unconcealment in which beings are. This change in Heidegger's conception of the clearing also entailed a metamorphosis in what it is to be (an entity): whereas before to be an entity was to be lit-up in understanding, now to be an entity is to be unconcealed.

A second change marking the advent of phase three is that Heidegger's attention shifted from being as the being of entities, entity-ness (*Seiendheit*), to being per se. In *Being and Time*, Heidegger still thought of being as the being of entities. Subsequently, Heidegger addressed being itself and claimed that this is what the history of philosophy since Plato had forgotten. For in interpreting being as the being of entities, this tradition had overlooked that, beyond the fact that entities are, being itself holds sway. "Holds sway" translates the verb *wesen* as Heidegger used it in phase three. The corresponding noun *Wesen*, though still translatable as "essence," no longer means, as it had in philosophical tradition, that which makes an entity the type of entity it is. Instead, *Wesen* means the happening that belongs to— or that marks—an entity as the entity it is: an entity's *Wesen* is how that entity comes to be and perdures as the entity it is. Heidegger corrected the tradition's oversight of being itself by, among other things, calling on the concept of the event (*Ereignis*). Whereas entityness is the mode of being characteristic of what shows up in the clearing, the concept of the event articulates the mode of being (if you will) that characterizes the clearing of being: whereas entities are, the clearing of being, the unconcealedness of entities, happens ("Es gibt sein," inadequately translated as "There is being"). Heidegger's transformation of the notion of *Wesen* from essence to happening that belongs to reflects his use of the concept of event to characterize being per se. (Of course, the event of the clearing of being must be distinguished from the myriad happenings standardly labeled "events," all of which occur *within* the clearing.) Lest talk, including his own talk, of being per se reify being as an entity, Heidegger emphasized the task of thinking the "ontological difference" between being and beings; he also sought to counter the inevitable hypostatization of being in language through the appropriation of archaic words (e.g., "Seyn"), new uses of familiar words, and orthographic innovations (such as "S✖n"). In sum, the transition from phase two to phase three of Heidegger's thought is the transition from the meaning of the being of entities to the truth of being as such. The third phase of Heidegger's philosophy can be thought of as a series of attempts to prepare thought for experiencing the truth of being.

A third significant change in phase three concerns the relation between humans and the clearing. In phase two, human being and the clearing were in some sense the same (though not identical). In phase three, they came apart: the clearing is distinct from human being. Humans understand, while the clearing happens. Humans now *stand into* the clearing that happens: the human understanding (or temporality) that earlier was taken to open up the clearing is now construed as that by virtue of which humans stand into it. As a result, whereas before Heidegger specified human being as existence (*Existenz*), he now characterized it as ek-sistence (*Ek-sistenz*) to emphasize this being opened up to. The clearing, however, cannot happen without humans standing into it. The term *Ereignis* (event) even suggests this because, read as *Er-eignis*, it suggests being appropriated by the happening of being. The indispensability of human existence for the event of being indicates that the open of unconcealedness happens to human being. In this sense, "man is the shepherd of being."[1c]

---

[1]   Martin Heidegger, "Letter on Humanism," in *Basic Writings*, David Farrel Krell (ed), London, Routledge and Kegan Paul, 1978, pp. 189-242, here p. 210;

Heidegger first conceptualized this standing into through the notion of understanding. Quickly, however, language came to play this
function: it is by virtue of language that humans stand into the clearing of being. "Language is the house of being," Heidegger famously
wrote in the "Letter on Humanism."[2,d] The clearing happens as articulated in language. This idea can be understood as follows. Suppose a
child is sitting at a table. The state of affairs involved in this supposition is that the child is sitting at the table. This state of affairs is the
particular state of affairs it is by virtue of the language articulating it,
i.e., "The child is sitting at the table." Other states of affairs, for example, that the child is sitting on a chair, are the states of affairs they are
because of the language that articulates them. Language, in other
words, is not merely a medium in which antecedently existing states of
affairs (or thoughts) are represented: states of affairs (facts) are not
independent of language. Rather, language articulates which state of
affairs any state of affairs is: it is by virtue of language that the state of
affairs, that the child is sitting at the table, is the state of affairs it is and
not some other one. A state of affairs, in short, is a linguistic entity,
although the child, the table, and sitting are not. (Moreover, its being
true that the child is sitting at the table depends on how things stand
with the child and the table: whether a given state of affairs—
articulated in language—obtains or not depends on the world.) In
short, to put things in Heideggerian language: language *discloses* the
world. It opens up possible states of affairs. In rephrased words of
Heidegger's contemporary Wittgenstein, "the limits of language are
the limits of the world."

I wrote above that much of Heidegger's later work consisted of interpretations of Western philosophers or Germanic poets. Heidegger
had conceived of the unbuilding (*Destruktion*) of the history of philosophy that was to be part three of *Being and Time* but was never
written as an undoing of the tradition based on an absolute account of
time and being. Later Heidegger, by contrast, often pursued the question of being through interpretations of history. The only extended
systematic account he composed (*Contributions to Philosophy*) was
kept private and published posthumously in 1989 only after all his
lectures series had been published. Many, moreover, of the general or

---

"Brief über den Humanismus," in *Wegmarken*, Frankfurt am Main, Klostermann, 1967, pp. 311-60, here p. 328.

[2]   Ibid, p. 213 (German p. 330).

systematic claims Heidegger made about being are based on his interpretations of past philosophers and poets.

In Heidegger's eyes, the history of the truth of being has three
overall phases: the prePlatonic, the metaphysical, and the postmetaphysical. Whereas Plato effected the transition from the first phase to
the metaphysical one, Nietzsche and the age of modern technology
completed the metaphysical phase. Hölderlin and Heidegger, moreover, together annunciated the possibility of a transition from the
metaphysical to a postmetaphysical world.

The prePlatonic era embraced a short period of time when being
showed itself to thinkers and poets. The two most prominent
philosophers whom Heidegger honored in this regard are Parmenides
and Heraclitus. What distinguishes them is that being as such—the
clearing and unconcealedness—was revealed to them. Accordingly,
Heidegger thought, existing fragments of their writings contain
insights of potential edificational value for contemporary humanity.
(Although Heidegger valorized these thinkers and in his thought
created an idealized prePlatonic Greek world, the ethics he advocated
in his later work never involved returning to a prePlatonic life).
Parmenides, for example, is said to be the first thinker to think being
as a whole as such. He and Heraclitus are equally praised for
understanding truth as unconcealedness, while Heraclitus, because of
his use of "phusis," also wins favor for grasping that concealment
(*Verbergung*) belongs to unconcealment (*Unverborgenheit*).
(*Unverborgenheit*, Heidegger argued, is clearing concealment [*lichtend
Verbergung*], an amalgam of uncovering [*Entbergen*] and concealing
[*Verbergen*]). Heidegger, finally, interpreted Parmenides' thesis that
man is the measure of all things as meaning that it is essential to the
happening of the clearing that humanity stand into it: the event of
being necessarily befalls humans. Another significant feature of
Heidegger's interpretation of the prePlatonic era is his juxtaposition of
poets such as Sophocles alongside philosophers as co-institutors and –
preservers of the happening truth of being.

The history of being began with the prePlatonics. After the
prePlatonics, the clearing, i.e., being as such, withdrew. Beginning
with Plato, being was understood as a particular feature of entities,
namely, presence (*Anwesenheit*): to be is to be present. The subsequent history of philosophy became a series of reconceptualizations of
presence, and being became fixed as the beingness of beings, the
entity-ness of entities (*Seiendheit der Seienden*). Plato also invented

the idea of an immutable realm different and higher than the protean realm of sense experience. He argued that sense objects are what they are by virtue of participating in the insubstantial forms (eidien) that populate this higher immutable realm. He thereby instituted ontotheology, which both defines as highest those entities that are most present (e.g., the forms or the Christian God) and grounds the unconcealedness of all other entities in these highest one(s).

Thus began the era of metaphysics, which is reaching its completion only today. The history of philosophy is largely the history of metaphysics. As just outlined, metaphysics, according to Heidegger, possesses three defining tenets: (1) that being is the entity-ness of entities as a whole, thus a feature of entities, thus itself something that is, (2) that this feature is presence, and (3) that higher than the sense world is the immutable world that contains the entities of the greatest presence that ground things of the sense world. The history of metaphysics is a series of reinterpretations of presence as the entity-ness of entities. Thereby forgotten is what Heraclitus and Parmenides had seen, namely, being, or presencing, as such.

The history of metaphysics is a series of epochs. Each epoch is the reign of a particular metaphysical understanding of the being of entities. This understanding is expressed in a key philosophical word and initially takes the form of the thought of a particular philosopher. Examples of these epochs are *eidos* (Plato), *ousia* (Aristotle), *energeia* (Aristotle), *actualitas* (Aquinas), *substantia* (Descartes), *perceptio* (Berkeley), *monad* (Leibniz), *Gegenständlichkeit* (objectness, Kant), and *Wille zur Wille* (will to will, Nietzsche). These thoughts are sent, or fated (*Geschicke*). Philosophers do not invent, or willfully excogitate, them. Since an epoch's instituting thought is simply given, any epoch is discontinuous with all previous epochs (though each instituting thought is a further specification of presence). An epoch just begins, the instituting thought subsequently spreading to the thinking of others. The instituting thought also disseminates through culture more widely. Heidegger believed that the history of metaphysics underlay the history of the West, though he never tried to make good on this proposition apart from his discussion of the contemporary world.

According to Heidegger, metaphysics is coming to completion. Nietzsche, he claimed, is the last metaphysician. Although Nietzsche spurned higher immutable realms and claimed that the world of sense experience is all there is (for Heidegger, Nietzsche's celebrated dictum that God is dead means that the era in which the presence of beings is grounded in a highest being has ended), he still thought being as the presence of entities: being is will to power, which Heidegger resolved into will to will. What is more, the cure that Nietzsche advocated for the existential malaise resulting from the conjunction of the death of God with nihilism (the reevaluation of hitherto highest values such as truth and the good) is a type of heroic willing (the overman [*Über-mensch*]). In thus instantiating a prime feature of the being of entities as it had been conceived of since Leibniz, namely, will, this prescription remains in the thralls of metaphysics. Heidegger also claimed, however, that with Nietzsche metaphysics had been philosophically completed: no more possibilities for understanding being as presence exist. It should be added that in the 1930's Heidegger appropriated the term "nihilism" from Nietzsche for the purposes of characterizing the condition of humanity. Whereas Nietzsche, however, had used the term to specify the plight of a Western world that, through its pursuit of truth, was losing belief in its hitherto highest values, Heidegger used it to name the fact that the 2300 year old metaphysical Western world had forgotten about being as such: nihilism is the absence/forgetting of being.

Heidegger further claimed that, whereas Nietzsche's thought instituted the final epoch of metaphysics, the 20th-century world realized this epoch in actual human life: the development of systematic technological societies is the concrete sociohistorical form assumed by the completion of metaphysics. The reign of modern technology involves a setting upon and setting up (*Bestellen*) of the world and the entities in it so that they are on call for the satisfaction of human wants and needs. Different from, say, the water wheel of the Medieval mill, which joined with the flows of nature to generate useful movement without substantially altering those flows, the modern hydroelectric dam so shapes and reorganizes—sets upon—aquatic nature that it can be commanded efficiently to yield useful products whenever human desire or need demands them. Nature becomes, in Heidegger's memorable phrase, standing reserve (*Bestand*). Together, setting upon/up and standing reserve constitute enframing (*Gestell*), the current clearing of being. As the ordering of the world as standing reserve becomes more systematic, human beings, too, increasingly become standing reserve, for instance, labor power at the disposal of capitalists. Heidegger feared that the Western world was developing into a total system in which everything would have its place as stand-ing reserve for the efficient satisfaction of whatever desires and needs are to be satisfied. Worse, the inner truth of enframing is that setting upon and setting up are ultimately carried out for a single end, namely, having everything under control for its own sake. Notice that the problem that this overall situation poses lies not with technology per se, but with the wider context in which modern technology is used. Heidegger was no Luddite.

Social systems represent the concrete sociohistorical completion of metaphysics because their emergence realizes Nietzsche's dictum that being is the will to power. Setting upon/up is the subjection of things to human molding for human benefit, thus their subjection to the human will to power. As the systemization of humans and the things with which they traffic so progresses that the end of control for its own sake dominates the ordering process, enframing becomes a realization of the will to will. Amid all this ordering, moreover, the contemporary world has forgotten the question of being. More distressingly, from Heidegger's perspective, the contemporary world, focused on effi-ciency, is emphatically antimetaphysical in spirit. If today the ques-tion is posed, "What is the use of metaphysics?," the answer that dominates is "Nothing." From Plato to Nietzsche, Heidegger argued, metaphysicians had been concerned with being. Although they had gone astray in theorizing a general feature of entities while neglecting the clearing of being, at least they were concerned with being. In the age of modern technology, philosophers lose their interest in being. They forsake, not just the clearing, but the entity-ness of entities, too, and thereby forget metaphysics' forgetfulness of being: the era of tech-nology is the oblivion of the oblivion of being as such. Indeed, whereas Nietzsche finished metaphysics in being its last thinker, the world of modern technology finishes off metaphysics in dispensing with it.

Is humanity condemned to be forsaken by being and to be sub-jected to increasing systemization? No, replied Heidegger. In this context, he quoted Hölderlin, his favorite poet, "But where the danger is, grows the saving power also." What Heidegger (not Hölderlin) meant is that today (the 1950s), with the completion of metaphysics in enframing, the possibility arises of reversing the forgetfulness that has characterized metaphysics and of putting Western humanity on a new ontological footing. The magnitude of the task must not be underestimated: Heidegger aimed at a language, a way of thinking, and a form of life that would break with the 2300 year metaphysical tradition of Western culture. One wonders whether a break with something so encompassing is possible. Heidegger, however, thought that several paths at least point in that direction.

One path is pondering the work of the prePlatonics, the thinkers who experienced and said what subsequent philosophy forgot. Pondering fragments of text in which being and the clearing show themselves prepares thought for a postmetaphysical era. A second path is examining the history of metaphysics—understanding its character, showing where it went astray, and appreciating its completion. The fact that enframing spurns metaphysics facilitates this path, for pondering how metaphysics has forgotten what needs to be thought likewise prepares thought for a postmetaphysical world. Many decades before Nietzsche, finally, lived a poet, Hölderlin, who not only anticipated Heidegger's understanding of metaphysics but also pointed toward the possibility of solutions to the predicament that humanity confronts at the completion of metaphysics. In Heidegger's eyes, Hölderlin is the herald of a postmetaphysical era. "What is essential is that we are caught up in the consummation of nihilism, that God is 'dead,' and every time-space for the godhead is covered up. That the surmounting of nihilism nevertheless announces itself in German poetic thinking and singing...."[3,e] Elucidating Hölderlin's poetry is a third path that promotes the future of Western humanity.

Hölderlin, Heidegger wrote, is the poet in needy times. To begin with, he claimed, Hölderlin understood metaphysics. Heidegger read Hölderlin's mourning for the withdrawal of the gods in the poem "Germania," for instance, as recognition that humanity has been forsaken by being, i.e., that metaphysics has forgotten the question of being. He similarly interpreted Hölderlin's interest in the Greeks as recognition that something was experienced in the prePlatonics that is worthy of thought today. More importantly, Hölderlin's poetry pointed toward the possibility of postmetaphysical existence while also

---

[3]   Martin Heidegger, "The Rectorate 1933/34: Facts and Thoughts," *Review of Metaphysics* 38 (1985): 481-502, here p. 498; "Das Rektorat 1933/34: Tatsachen und Gedanken," in *Die Selbstbehauptung der deutschen Universität: Rede, gehalten bei der feierlichen Übernahme des Rektorats der Universität Freiburg i. Br. am 27.5.1933. Das Rektorat 1933/34: Tatsachen und Gedanken*, Vittorio Klostermann, Frankfurt a/M, 1983, p. 39.

offering concepts with which the outlines of such an existence could be comprehended.

Above all in the 1950s, Heidegger appropriated concepts from Hölderlin for the purpose of comprehending the clearing and human existence. He reconceptualized the happening of the clearing of being in Hölderlinian terms as the mirror-play (*Spiegel-Spiel*) of the oneness (*Einfalt*) of earth, sky, mortals (*Sterblichen*), and divinities (*Göttlichen*). Earth and sky are, respectively, that on and under which mortals, i.e., humans exist (humans are the mortals because humans alone are capable of death as death). The divinities, meanwhile, are the measures for human life, the granters of well-being (*Heilgewährenden*). The happening of unconcealedness is the mirror-play of this fourfold, where the clearing is the open between earth and sky, the mortals are whom this open between befalls, and the divinities are the measure of the open. As will be described in chapter four, moreover, this mirror-play is admitted (*eingeräumt*), gathered (*gesammelt*), and preserved (*geschont*) in the thinging (*Dingen*) of the things (*Dinge*) with-at which humans sojourn (*sich aufhalten bei*). These things thereby constitute sites (*Stätten*) at which the clearing happens. When things thing, the fourfold are brought together out of their farness, resulting in a local unified situation of human dwelling.

This conceptual apparatus describes the clearing of being. It also specifies, in formal and general terms, a way of life commensurate with the structure of the clearing, namely, living appropriately to the role allotted to humans in the mirror-play. The role allotted to humans is that of being the shepherd of being, the entity that, in its language, poetry, and activity, gathers what is and preserves it in its being. As Heidegger put it in the *Contributions to Philosophy* using language carried over from *Being and Time*, the role that falls to humans is to be the there, Da-sein. A key component of a life commensurate with this gathering and preserving is letting the components of the fourfold be what they are: letting earth, sky, mortals, and divinities be what they are when the thinging of things draws them into oneness, and caring for them in their being. This letting-be is a key component of the ethics that Heidegger advocated in the third phase of his career. In, however, the age of technology, the age of setting upon/up and standing reserve, humans have forsaken this way of being: they do not let things be. Although setting upon/up is a type of uncovering, and although enframing is a type of mirror-play of the fourfold, humanity has forsaken its proper role of gathering and preserving. Notice, incidentally, that clearings are multiple. Different groups of mortals stand into different clearings. In these local clearings, moreover, the four components of the fourfold, including the divinities and the way of life of mortals (whether or not this way of life lives up to the role alloted to mortals in the mirror-play), assume specific forms. This particularization implies that the clearings into which different groups of mortals stand have different contents.

Heidegger not only appropriated Hölderlin's concepts for the purposes of conceptualizing the clearing and of formulating the outlines of a form of life different from setting upon/up. He also adopted Hölderlin's idea that poetry plays a special role in the historical institution of clearings. Poetry, Hölderlin wrote, responds to the wink-nod (*Wink*) of the divinities. Poeticizing (*Dichten*) is the naming of the holy (*Nennen des Heilegen*), the taking of the measure (*Nehmen des Masses*), the bringing of the signs of the divinities—equivalent ways of saying that poetry institutes the basic orientations and sensibilities according to which people live in a given clearing: poetry measures out, or opens up, the between of earth and sky where humans live. Hölderlin's language and ideas also specify a crucial, but still unrealized, task, namely, awaiting and responding to a new wink-nod of the divinities. Part of the cure to the contemporary homelessness of humanity—humanity's forsakenness by being and its straying from its proper role—is attentive awaiting for a new measure for human habitation. This means that preparatory to a life commensurate with the role allotted to humans in the mirror-play are attentive contemplative thinking (*Andenken*, the title of another Hölderlin poem) and an openness to the sending of a new nonmetaphysical understanding of being. These, too, are part of Heidegger's postmetaphysical ethics. In the age of technology, however, humans have neglected these ways of being. Yet, in conveying these matters Hölderlin's poetry, which speaks of the homecoming (cf. the poem "Homecoming"), is itself the homecoming of Western humanity to its role as shepherd of being: "…in instituting the essence of poetry anew, Hölderlin first determines a new time. It is the time of the gods who have fled and of the god who is coming."[4,f]

---

[4]   Martin Heidegger, "Hölderlin and the Essence of Poetry," in *Elucidations of Hölderlin's Poetry*, trans. Keith Hoeller, Amherst, Humanity, 2000, pp. 51-66,

   The name that Heidegger gave to the way of life needed today is releasement, or letting-be (*Gelassenheit*). *Gelassenheit* is humanity living up to its role as gatherer and preserver of what is. It involves, first,
humans so acting toward earth, sky, each other, and the divinities that
these are allowed to be. It involves, second, attentive contemplative
thinking, whereby humans await a new sending (*Schickung*) of being.
This new dispensation can only be awaited. Humans can do nothing
to bring it about. At best, by pondering the prePlatonics, studying
Heidegger, and reading Hölderlin, they can attain a readiness for a
new fate (*Geschick*). In Hölderlin's terms, humanity must be open for
the return of the gods. In an interview published in *Der Spiegel* after
his death, Heidegger stated that "only a god can save us now." What
he meant is that humanity must be open to receive a new basic sensibility and orientation for living, and that the fate of the West, the possibility that it will escape the consummation of metaphysics in enframing, depends on this readiness.
   A more secular way of putting this claim is that poetry alone can
save us now—an authentic poetry that takes the measure according to
which Western humanity can live anew. Hölderlin's poetry does not
achieve this. His work describes the predicament faced by the Western world at the end of metaphysics and sketches what must occur:
how, formally speaking, human life must change and what, formally
speaking, the cure will consist in. It does not, however, describe the
specifics of postmetaphysical life. For this, humanity needs new works
of poetry, language, and art, new works that lay down (*durchmisst*)
new clearings as receptacles of dispensations different from setting
upon/up and standing reserve. At the end of his essay on technology,
Heidegger questions whether it is possible for poetry—understood as
including the plastic arts and architecture—to perform this function in
the age of modern technology. He was skeptical of the powers of non-
representational art in this regard. His attitude was that one must just
wait, in the meantime practicing and fostering *Gelassenheit* wherever
and whenever possible.

   here p. 64, translation amended; "Hölderlin und das Wesen der Dichtung," in
   Erläuterungen zur Hölderlins Dichtung, Frankfurt am Main, Klostermann,
   1971, pp. 33-48, here p. 47.

4   Space, Spatiality, and Society

Space plays an uncertain role in the thought of Martin Heidegger. Although his account, in *Being and Time*, of the spatiality of lived experience has inspired subsequent phenomenological analyses of human spatiality, this account is relatively short and plays no obvious systematic role in Heidegger's magnum opus. In that book, temporality is the meaning of Dasein's being, time is the meaning of being, and spatiality (*Räumlichkeit*) makes but a single more than cursory appearance after the careful analysis of it in the first half of the book—only to be analyzed as a mode of temporality. In his third phase, moreover, Heidegger regularly wrote of the place (*Ort*) where the clearing occurs, but he rarely explored the spatial nature of this place. What he had to say is tantalizing, and space now assumes a systematic position in the truth of being, but space and its relations to truth, being, and time remain underdeveloped themes. In short, although Heidegger is a major theorist of space, he wrote remarkably little on the topic. Regardless of his readers' opinions of his leading issue, the question of being, Heidegger's remarks on space exemplify how his thought roamed across a remarkable range of topics, leaving behind insights on numerous matters.

Categories of Space

It will help clarify Heidegger's thoughts on space if the types of space he analyzed are contrasted with two basic categories of space: objective and subjective.

Objective space is space conceived of as an objective feature of the world. An objective feature of the world is one that persists independently of human apprehension, comprehension, and activity. There are two basic conceptions of objective space, absolute and relational. Objective space conceived of as absolute is a thing unto itself. More specifically, it is a container, or arena, in which events occur and objects exist. To claim that objective space is a thing unto itself is to hold that this container or arena exists even when no events or objects populate it. Contemporary astronomy reports, for example, that relatively few entities exist in the vast expanses between galaxies. According to an absolute conception of space, say, as a container comprising all possible points where entities can be, space exists even in these immense interstices as a manifold of possible places for entities. Other features of absolute space are homogeneity, the fact that the properties of space are evenly distributed throughout it, and isotropy, the fact that it lacks inherent directions or surfaces.

The second basic conception of objective space is relational space. Although conceptions of relational space are conceptions of a space that perdures independently of human minds and actions, these conceptions do not construe space as independent of objects. According to these conceptions, it is not the case that objects have spatial properties by virtue of existing at positions contained in a subsisting container or arena. Rather, objects possess spatial properties as and by virtue of relations to one another: space is a collection of relations among objects and of properties based on these relations. Space is unimaginable, as a result, in the absence of objects. This fact does not imply, however, that relational objective space is absent, say, in between galaxies. For the few entities that do exist there stand in spatial relations to one another and to other existing objects, for example, nearby galaxies and the black holes, stars, planets, and dust that compose them. Relational space, finally, is neither homogeneous nor isotropic. Spatial relations and features can vary greatly from entity to entity, and most arrangements of relations and features are oriented.

I should add that relativistic space is a possible category of objective space. Conceiving of space as relativistic means treating the spatial positions of events and objects as relative to frame of reference. When frames of reference are something other than human experience, space so conceived is objective. When, however, frames of reference are human experiences, i.e., observations of events and measurements of spatial positions, relativistic space is not objective.

Conceptions of absolute and relational space have been prominent in modern philosophy and physical theory. As Heidegger remarked,[1] labeling them "objective" reflects the "Cartesian" contrast between conscious subjects and worldly objects that dominated the modern era. This subject-object split also made the notion of subjective space possible. The mark of space conceived of as subjective is that it depends, not on the objects, but on the subjects of experience. One very famous conception of space as subjective is Kant's thesis that space is an a priori form of intuition. What Kant meant is that, although what stands in spatial relations and possesses spatial properties are the objects, and not the subjects, of experience, objects stand in these relations and possess these properties because the subject imposes them on the (mental) representations that make up its experience of objects (for Kant objects are representations). Spatial relations and properties are a priori: they characterize any possible object of (outer) experience because the subject imposes them on all its experiences regardless of the latter's content. Space is also a form of intuition: the representations upon which the relations and properties in question are imposed—as constitutive of these representations—are called "intuitions." Incidentally, the supposition that space is subjective does not entail that reality is subjective. It implies only that any reality that is independent of subjects is not in space and has no spatial properties.

So long as the subject-object polarity dominated philosophical thought, philosophical discussions of space were tied to the contrast between subjective and objective. Heidegger's analyses of space are of a third category of space, lived space. The expression "lived space" has multiple meanings. It can mean, first, the space, possibly objective space, that people experience as they go about their business. Here, "lived" denotes those aspects of the world that enter the ken of ongoing experience. "Lived space" can mean, second, the space that belongs to experience itself, where experience is construed as something distinct, though not necessarily independent or separate, from the objects of experience (e.g., the seeing of the landscape as opposed to the landscape seen). When "lived space" is so interpreted, it embraces features of experience that might not characterize the objects of ex-

---

[1]   Martin   Heidegger, "Art and Space," trans. Charles H. Seibert, *Man and World* 3 (1973): 3-8, here p. 4; *Die Kunst und der Raum*, St. Gallon, Erker, 1969, p. 7.

perience. The point of calling such features "lived" (instead of, say, "experienced") is to stress the active dimensions of people's engagement with the world. "Lived space" can mean, third, the space of lived experience, that is, the space that characterizes the phenomenon of humans experientially acting. By "experientially acting" I mean that people experience their carrying on in the world and that a person's experience occurs within the ken of his or her activity. Unlike experience in the case of the second type of lived space, the phenomenon of experientially acting is not distinct from the objects and events with which a person deals. The space of experientially acting is a property neither of these objects and events as opposed to experientially acting nor of experientially acting as opposed to these objects and events. This space is, instead, a property of the complex phenomenon: carrying on with and amid these objects and events. "Lived space," finally, can mean the space of living, the spaces with which human life is involved, in terms of which it proceeds. An example of such a space is a group's territory. It is not necessary that the spaces involved are experienced, or even cognized, as spaces with which humans are involved, though either might be true.

Heidegger analyzed different types of lived space in phases two and three of his career. Whereas the lived space examined in phase two is the space of acting experientially (type three), the lived spaces theorized in phase three are spaces with which human life is involved (type four). Lived spaces of both sorts differ from objective spaces because, even though they, like objective spaces, encompass objects in the world, they do not, as do objective spaces, characterize the world independently of humans. The space of experientially acting and the spaces with which humans are involved are characteristics of the world only insofar as humans proceed in them; what's more, their characters depend on features of human life. As we shall see, both sorts of lived space also differ from subjective space because neither depends on minds or subjects of experience. Heidegger's lived spaces have nothing to do with minds or subjects of experience because, as discussed in the previous chapter, Heidegger's thesis that human existence is being-in-the-world contests the idea that people are either minds that represent the world or subjects who stand over against a world of objects—and Heidegger's later thought never abandoned this position. The character of lived space does, to be sure, depend on features of human life, but these are features of activity and not of mind or subjectivity: both sorts of lived space are nonobjective, non-subjective spaces of activity. The difference between subjective space and these lived spaces is the difference between a mind-imposed or subject-dependent space, on the one hand, and the space of being-in-the-world or of dwelling amid (*bei*) things on the other.

The Spatiality of Existence

Heidegger never used the expression "lived space" in *Being and Time*. Nevertheless, his account of the space of being-in-the-world is the first analysis ever given of the third type of lived space distinguished above. Heidegger's discussion of space in that book focuses almost entirely on this lived space. He acknowledged, however, that objective space is pertinent to human life. To mark the difference between lived and objective space, Heidegger usually reserved the word "spatiality" (*Räumlichkeit*) for the space of being-in-the-world. He used the word "space" (*Raum*), however, for both the space of existence and objective space, a duplexity that can cause interpretive confusion. Heidegger did not have a word for subjective space. Since Heidegger did not believe that human beings are subjects confronting objects, he did not believe in subjective space.

Explaining Heidegger's account of the spatiality of human existence requires first considering his analysis of world and worldhood (*Weltlichkeit*). In the previous chapter, I explained Heidegger's thesis that the practical mode of being-in (-the-world) has priority over the cognitive mode of being-in. I also indicated that, in accordance with this priority, Heidegger conceptualized the entities that make up a world as a nexus of handy equipment, as opposed to a collection of extended things. Heidegger contended, further, that what is responsible for a collection of equipment forming a nexus is a complex of possible ends, projects, and uses. Consider, for example, the facts that equipment are characterized by uses to which they can be put and that they are usually defined by some subset of these. A hammer, for example, can be used to fasten together boards, to loosen annealed pieces of metal, and to prop open a window, and it is defined (as hammer) by those uses that fall under the heading "hammering." Hammers are put to uses such as these because of the ends people pursue and the projects they carry out in pursuing them. For example, wanting her children not to watch too much TV (an end), a woman might use a hammer to fasten together boards with nails (a use) as part of the project of building a playhouse for them. The particular uses that characterize equipment thus reflect people's ends and projects. People can, of course, pursue multiple ends and carry out various projects for the sake of their ends. As a result, equipment are characterized by a range of *possible* uses that devolve from the complex of ends and projects that people might take up. Heidegger called such a complex of possible ends and projects (which are, respectively, states of being and actions) significance (*Bedeutsamkeit*). An important fact about the elements of any whole of significance is that they essentially relate. So, too, consequently, do the equipment whose meaningfulness as equipment is defined by that significance. Equipment thus form nexuses, as opposed to heaps or mere collections. In a work room, for instance, hammers, saws, nails, workbenches, wood, tape measures, and the like interrelate by virtue of the varied and interrelated uses that can be made of them. The varied and interrelated uses that can be made of them are, more fully, uses of them in certain workroom projects that are carried out for the sake of particular ends.

A world is "that, '*wherein*' a factical Dasein as such 'lives.'"[a] (SZ 65) The world wherein someone lives when proceeding in the workroom comprises the equipment found there (and possibly elsewhere). The structure of that world, moreover, is that complex of possible ends, projects, and uses (ways of being and actions) by virtue of which the equipment involved form a meaningful nexus. The world wherein a person lives at any moment, consequently, is a nexus of teleologically meaningful entities. What it is for a person to carry on in that world is for that person to use that world's equipment in projects and for ends that are contained in the whole of significance that organizes this equipment as a nexus: being-in-the-world is proceeding amid equipment according to the teleological significance that characterizes them.

The nexus of equipment amid which a person proceeds at any moment forms a world-around (*Umwelt*). It is the world "around"— amid, with, and toward whose entities that person acts. Heidegger claimed that the around (*Um*) of the world-around makes up the world's spatiality. The spatiality of the world is the world's being around, and not over against, the person in it; the person's being amid (*bei*), and not, as in the Cartesian picture of a subject confronting a world of objects, over against (*gegenüber*), the entities that compose it. "We shall designate the phenomenal structure of the worldhood of space as the *aroundness* of the world as *world-around*."[b] (PHT 224, translation amended).

Heidegger's discussion of the spatiality of being-in-the-world is divided into two parts, one about the spatiality of the entities that compose the world and one about the spatiality of being-in. These two discussions strongly overlap. Each is a discussion of the spatiality of existence, the spatiality of being-in-the-world, that emphasizes particular features of this spatiality while referring, implicitly or explicitly, to the entire phenomenon. That these two discussions overlap reflects the fact that being-in-the-world is a unified phenomenon that cannot be sundered into two components, being-in and world, which division might be comparable to the subject-object split of modern philosophy. The spatiality of being-in and the spatiality of the world are overlapping combinations of the phenomena that compose the space of human life, the spatiality of being-in-the-world.

The spatiality of the nexus of equipment that composes a world has two fundamental clusters of features: nearness and farness, and place and region. Heidegger characterized the being of equipment as readiness-to-hand, or better, handiness (*Zuhandenheit*), which can be interpreted as availability for and involvement in people's activities. The word itself directly evokes nearness: *Zuhandenheit* literally means being at hand (*zu Hand*), that is, near. In the case of equipment, however, being near, or at hand, is not the same as lying at a relatively short objective distance (*Abstand*) from the person acting. Rather, near means right at hand, right about, available for use or actually being used. This is not the same as objective nearness. Something can be used or available for use even though it lies at a relatively large objective distance from the person concerned. An example is a school computer network that a professor or student accesses from thousands of miles away. Conversely, entities that are objectively close can be unavailable for use, for example, molecules on the surface of the actor's skin. Entities can also be more or less available in or for use, more or less near or far vis-à-vis ongoing activity. In general, the equipment that make up a world lie at varying nears or fars from the people who are-in-that-world, that is, who are acting in it. Indeed, for a piece of equipment to belong to a world is for it to fall somewhere in the near-far spatiality spectrum for any person who is in that world.

An entity, "so far as it is there within the world, is in general [*über-haupt*] distant from me, in general has a possible near to and far from me ..."[c] (PHT 225, translation amended).

Within a world any piece of equipment has a place (*Platz*). The place of a piece of equipment is its where (*Wo*). This where is the there (*Dort* or *Da*) of its belonging-to (*Hingehörens*). Heidegger had a very specific understanding of the belonging-to involved: "Belonging-ness-to in each case corresponds to the equipment character of what is handy..."[d] (SZ 102). Generally speaking, what any piece of equipment belongs to is human activity, and the to (*Hin*) of the belongingness-to (*Hingehörigkeit*) of a specific piece of equipment is the action it does or can subserve, the use to which it is or can be put. The to of a work-bench, for example, is building or repairing things. The place of a piece of equipment, the there of its belonging-to, is where it belongs in or fits into human activities. Heidegger wrote, accordingly, that the place (*Platz*) of a piece of equipment is the place of this equipment to X, where X is the action it subserves.

The notion of place has a peculiar complexity. On the one hand, equipment compose, *are*, places, namely, places where something can be done, places where specific activities can be performed: a work-bench is a place to build and fix things. On the other hand, equipment *have* places, they are placed (*platziert*): the place of the workbench is to be where people can build and fix things. This complexity is not an ambiguity. For the place that a piece of equipment is *is* the place that it possesses. A workbench is a place to build and repair, which is also its place in people's activity. Notice that place, a spatial property of equipment, depends on people, more specifically, on what they do and are up to in acting. This dependency illustrates the idea that being-in-the-world is a unified phenomenon.

Heidegger claimed that a place is always one place in a totality of places (*Platzganzheit*). This fact is tied to the interrelatedness of equipment, which itself devolves from the complex of possible ends, projects, and uses that people can pursue and carry out in the world constituted by these placed equipment. For example, the place of a workbench, to build or repair things, is tied to the place of a hammer, to attach and remove nails, both of which are tied to the places of boards, nails, saws, levels, vises, and so on. Equipment forms equip-mental wholes; correlatively, the places of equipment form place

wholes. As suggested, both wholes are tied up with the whole of significance that structures the world in question.

Heidegger called such a place whole a region (*Gegend*). The region
is the "where of the to-which of a belonging-to, going-to, bringing-to,
looking at, and the like."[e] (PHT 228, translation changed) It is where,
overall, the activity that uses equipment, and by reference to which
that equipment has places, is performed. The region is made up of a
totality of possible places as defined by the possible actions that a particular totality of equipment can subserve for certain possible ends.
Heidegger claimed that the region has priority over the places com-
posing it: "Something like a region must first be discovered if there is
to be any possibility of allotting or coming across places for a totality
of equipment that is circumspectively at one's disposal."[f] (SZ 103)
That is, Dasein must already be familiar with a totality of possible
places for equipment in order to assign or discover the specific places
of the equipment involved. Notice that he wrote that places can be
either assigned or discovered. Heidegger held that a person primarily
proceeds, not through places that she herself assigns to the entities
with which she deals, but through places that she discovers because
they are already assigned to them (by social norms, *das Man*—see
below). Heidegger also claimed, incidentally, that being familiar with
a region is a condition of spatially encountering entities: dealing with a
piece of equipment that occupies a place in some region presupposes
familiarity with the region. Picking up and tightening a screw with a
screwdriver, for example, presupposes that the person who does this
already understands some totality—however small—of placed equipment, of which the screwdriver and the screw are elements.

The region is tied to Dasein's significance, the whole of ends, projects, and uses that determines the meaningfulness of the current
world. The spatiality of a totality of equipment

> has its own unity...through the world-ish totality of involvements...The
> 'world-around' does not arrange itself in a space that has been given in
> advance, rather its specific worldhood articulates in its significance the
> context of involvements of any current totality of circumspectively allot-
> ted places. In each case the world discovers the spatiality of the space that
> belongs to it.[g] (SZ 104, translation modified)

Heidegger's example of the sun well illustrates the intertwining of
significance and region. The sun is an entity that is continually handy.

More specifically, its light and warmth (and the darkness and coldness of their absence) are used daily. This means that local worlds, for instance, a porch, a sunroom, or the distribution of rooms in a house, are laid out taking warmth/coldness and light/darkness into account. At the same time, the sun has its own significant phases, or celestial regions (*Himmelsgegenden*) defined by reference to its warmth and light: dawn, noon, dusk, and midnight. In taking account of warmth/coldness and light/darkness in laying out their local worlds, people arrange their worlds by reference to these celestial regions. For instance, fields, cemeteries, churches, and many military camps were and are still laid out according to the rising and setting of the sun.

The region also contributes to the aroundness of the world-around, the fact that a person always proceeds amid, and not over against, the entities that make up a world: "The regional orientation of the multiplicity of places that belongs to what is handy makes up the aroundness—the round-about-us—of those world-aroundish entities that we firstly encounter."[h] (SZ 103, translation amended) A person always finds herself amid the entities that compose a region. As discussed, moreover, these entities are differentially near and far according to their involvement in her activity. In sum, the around of the world-around is a region of interrelatedly placed equipment differentially near and far vis-à-vis activity. The around is the regionalized near and far of the entities amid which people concernfully act.

It is scarcely necessary to point out the contrast between this sort of space and objective space. In depending on human activity and the ends people pursue in their activities, the spatiality of the world hardly persists independently of human apprehension, comprehension, and activity. Similarly, the differential near and far that characterizes regionalized equipment is a matter of the involvement of these entities in activity. This nearness and farness, furthermore, cannot be measured by reference to repeatable units. At best one can distinguish nearer and farther. The same holds of regions and places themselves: they do not populate an objective arena or form a objective distribution that can be measured by a uniform metric or coordinate system. At best, one can distinguish between more numerous and fewer places and between richer and thinner regions.

As described, Heidegger's analysis of the spatiality of being-in-the-world has two parts. The first is his analysis of the spatiality of equipment or world. The second is his description of the spatiality of being-in, the spatiality of people's practical engagements in the world. As indicated, moreover, these two analyses overlap because being-in-the-world is a unified phenomenon. Indeed, Heidegger's analysis of the spatiality of being-in greatly repeats, with different emphases, his analysis of the spatiality of equipment.

Being-in (*In-sein*) is itself a spatial phenomenon. The sense, however, in which people are-in (-the-world) must be distinguished from the sense in which objects are in, or inside, one another. We say, for example, that water is in the glass and that the glass is in the refrigerator. This "in" signifies a relationship that extended objects have to one another with regard to their location in objective space (SZ 54), namely, the location (extension) of the one being entirely surrounded by the location (extension) of the other. A person is not "in" the world in this way. Being there is not a matter of, say, a person's body being entirely surrounded by the objects that make up her current world. The sense in which people are in the world is, instead, in-volvement: to be in the world is to be involved in it, to be dealing with and proceeding amid (*bei*) the entities that compose it. This "in", Heidegger wrote, stems from "'*innan*'—to dwell, *habitare*, sojourn. '*At*' signifies "I am accustomed." "I am familiar with," "I look after something"…'*I am*' signifies: I dwell, sojourn amid the world, as what is familiar."[i] (SZ 54, translation modified) To be in-the-world is concernfully to proceed amid worldly entities in carrying out one's projects for the sake of one's ends.

The spatiality of this concernful dealing with has two principal aspects: orientation (*Ausrichtung*) and dis-severance, or better, dis-tance (*Ent-fernung*). Both phenomena were encountered in the above discussion of the spatiality of equipment without being emphasized. Orientation is the orientation of being-in-the-world. This orientation lies in the equipment-using actions people perform, the uses people make of equipment. These uses are tied to the projects people carry out and the ends they pursue in doing so. Like the to (*Hin*) of the belongingness-to (*Hingehörigkeit*) of the equipment involved, this orientation is a teleological directedness ultimately rooted in ends (and also in *das Man*, see below). Indeed, the orientation of activity and the to that belongs to the equipment therein used are one and the same.

Orientation is also connected to the three-fold composition of being-in discussed in the previous chapter. Being-in, recall, is composed

of attunement, falling, and understanding, which Heidegger reinterpreted as the past, present, and future dimensions of the temporality
of human existence. Whereas attunement is how someone already is,
and understanding is the projection of possible ways of being, falling is
dealing with particular entities in carrying out particular projects in
pursuit of particular ends: a person "falls" into the world as doing such
and such and up to this and that. The particular actions that make up
falling are the orientation of being-in-the-world: Dasein's constant
directedness into the region around it, its constant having to do with
particular events and things as it carries on there. Since these actions
are embedded in projects and ends, the orientation of being-in-the-
world is teleological in character. I should add that this falling, this
directedness into the world, is thrown: a person always exists as
thrown into, that is, always already falling into the world. The sense of
movement, inherited from the late nineteenth century German philosopher, psychologist, and historian Wilhelm Dilthey, is palpable
here.

The second feature of the spatiality of being-in is dis-tance (*Ent-
fernung*). Heidegger inserted a hyphen into the common German
word *Entfernung* to signal that what he meant diverges from the
common meaning of the word. Whereas *Entfernung* normally means
distance, *Ent-fernung* means the elimination or overcoming of dis-
tance, i.e., bringing-near. As noted, the equipment that compose a
world are always differentially near and far vis-à-vis the activities carried on there; these nears and fars are thus relative to what people are
doing. Whereas a saw, for instance, is near in the activity of cutting
boards, it might be distant in the activity of correcting a crooked picture. As a person carries on in a region, his or her activity might so
change that equipment that was far is brought near. If the person
cutting boards suddenly notices a leaky pipe, a wrench that earlier was
far might suddenly be brought near. Indeed, as a person carries out
different activities over the course of the day, a great variety of equipment is brought near (or made to recede). As noted, the nears and
fars of equipment are distinct from the objective distances of occurrent things from other such things.

Heidegger distinguished three modes of dis-tance. The first is the
dis-tance of what is attended to. What someone attends to is what is
brought nearest in current activity. Coming near in the sense of being
attended to obviously has nothing to do with objective distance; a

sporting event on the other side of the globe might be what is nearest in this sense. The second mode of nearness is the nearness of being used in current activity. In this sense, the glasses the viewer wears when watching the sporting event have been brought near, unlike the scissors that lie unused next to the television (the event, however, has been brought nearer than the glasses have). The third mode of bringing-near is maintaining an "average field of reaching, grasping, and seeing" (SZ 106-7) in which entities are available. Something is at hand in a distinctive sense if it lies in this field, though it is not therewith brought as close as are entities that are attended to or used. Heidegger did not discuss a fourth mode of bringing-close: relevance. Equipment is brought near when it becomes relevant to ongoing activity. The wrench that becomes relevant when the leaky pipe is noticed is an example.

As a person proceeds, the entities in the region about her are differentially brought near and made far vis-à-vis her activity. Indeed, every entity that belongs to a given world lies near or far vis-à-vis the activities carried out there. As people go about their business, moreover, the nears and fars of the equipment about them metamorphose. "Because," Heidegger wrote, "Dasein is essentially spatial in the way of dis-tance, its dealings always keep in a world-around that is distant from it in a certain leeway."ʲ (SZ 107, translation modified) The distance of the world-around from a person is the nears and fars of equipment relative to her activity, and there is leeway for changes in their nearness and farness corresponding to changes in activity. This leeway thus depends on the possible activities of the people involved. It also depends, as we shall soon see, on the normativity of *das Man*.

The pattern of near and far also defines a person's "here." Where someone is is not defined by the physical location (*Stelle*) of his body. Rather, his "here" is an "amid-which of a dis-tancing being amid…in one with this dis-tance."ᵏ (SZ 107, translation altered). A person's here is defined by the pattern of near and far that helps compose the aroundness of the entities amid which he is carrying on, above all, those which he is using or attending to: it is a here having to do with certain entities bringing some near and making others far in doing such and such for the purpose of so and so. The there (*Dort*) that is coordinated with this here encompasses the places of near and far equipment. The here of a person's situation, meanwhile, embraces the leeway of the nears and fars of equipment in ongoing activity: here with different possible nears and fars tied to possible changes in activity. The there (*Dort*) coordinated with the here of the situation encompasses the possible places of near and far equipment. This there is the region.

Heidegger's insistence on keeping being-in free of extension and physicality led him to attribute the distinction between left and right to the orientation and dis-tance of being-in. Philosophers have traditionally attributed this distinction to physical structures of the human body, for example, the stereoscopic character of vision, the asymmetries of opposite sides of the body, or oppositely directed bodily movement. The human body, however, plays little role in Heidegger's analysis of being-in-the-world. In cursory remarks about left and right, he claimed that left and right are directions (*Richtungen*) of orientedness in a handy world. This means that to the left, to the right, and straight-ahead are permanent possibilities for a person's constant dealing with—constant grasping, moving, and looking at—entities. Because of this, the entities that are presently nearest in ongoing activity have the permanent possibilities of being to the left or right of those which either just were or next are nearest. Acknowledging that people are bodily (*leibhaft*), Heidegger went on to claim that it is because left and right are permanent possibilities of a person's directedness into the world that bodiliness (*Leiblichkeit*), too, boasts a left and right; this is why, among other things, people have left and right hands, arms, feet, and legs. Heidegger thus held that certain features of anatomical structure depend on the existential structure of people's directness into the world. It is hard to know what to make of this claim. Is it a causal claim about the evolution of bodily form? Is it a transcendental thesis about the condition of the possibility of certain facts about the world? Is it a deduction from phenomenologically ascertained facts? What's more, Just how does the phenomenon of being-in-the-world relate to the human body?, and What kind of body is at issue here—a physical one or an existential one? Heidegger's scant remarks leave these questions hanging. His account of spatiality thus suffers from a lack of clarity about how embodiment and the human body are bound up with spatiality. His phenomenological successor, Maurice Merleau-Ponty, more directly and successfully addressed the issues that arise here.

I noted that Heidegger believed that, not lived space alone, but objective space, too, bears on human existence. When entities are encountered in the cognitive mode of being-in, they are encountered as present-at-hand (*vorhandenen*), that is, occurrent (*vorkommenden*) things. One way of encountering entities cognitively is through a breakdown in successful ongoing activity. When, say, the head of the hammer I am using flies off, leaving me staring at the headless stem in my hand, the stem is encountered as an occurrent thing, in the way it looks and feels. It is not encountered as relating to the ends, projects, and uses relative to which the hammer had been something available for and involved in my activity. When what were handy entities are encountered as occurrent objects via a breakdown in activity, they are said to be de-worlded (*entweltet*). The regionalized places these entities formally occupied within the world drop away, and they are now merely occurrent things existing at points in objective space (whether conceived of absolutely or relationally).

> Places—and indeed the whole circumspectively oriented totality of places belonging to equipment ready-to-hand—sink to a multiplicity of positions for arbitrary things. The spatiality of what is ready-to-hand within the world loses its involvement-character…The world loses is specific aroundness; the world-around becomes the world of nature. The 'world' as totality of equipment ready-to-hand becomes spatialized to a context of extended things which are just present-at-hand and no more.[1] (SZ 112, translation slightly modified)

Breakdowns, of course, occur. Moreover, as Heidegger eventually acknowledged, people can assume the cognitive mode of being-in and encounter present-at-hand things in objective space without undergoing a breakdown and even while continuing to use equipment in the practical mode of being-in. However present-at-hand things come to be encountered, furthermore, they are encountered as occupying locations in objective space and not in terms of the significance structure of the world, thus not as regionally placed. In short, humans manage objective space, just as they cope with regionalized places. Like occurrent things, furthermore, the objective space that occurrent things occupy can be scrutinized and measured with fixed metrics. It can also be thematized and thereby become a topic for mathematics, architecture, and the physical sciences.

In the practical mode of being-in, the space of the equipment a person deals with is composed of near and far regional placement. In the cognitive mode of being-in, the space of the occurrent things that a person looks at, thematizes, and theorizes, embraces a manifold of points, distances, directions, and other objective properties either absolutely, relationally, or relativistically defined. Heidegger had a somewhat dichotomous sense of these two modes. At least most of his formulations suggest that a person is exclusively in one or the other. At the same time, Heidegger observed that when people encounter occurrent things, these things show themselves as already having been there—for example, when one had been using equipment. That is, the handy entities a person uses while in the practical mode are also all along extended things, though not encountered as such. It follows that objective space is always present as a person orientedly and dis-stancingly proceeds amid regionalized places. It is there, covered up (*verborgen*).

The omnipresence of objective space does not imply that objective space somehow grounds regionalized places or is prior in some sense to the space of existence. Heidegger claimed that theorists have erred in trying to piece together the handy properties of equipment out of occurrent properties of things and also that presence-at-hand (*Vor-handenheit*) cannot make readiness-to-hand (*Zuhandenheit*) intelligi-ble. Moreover, just as the practical mode of being-in has priority over the cognitive mode, he claimed that the space of equipment has prior-ity over objective space in that (1) near-far regional placement charac-terizes the entities that humans firstly and almost automatically en-counter when existing and (2) a person is able to encounter things in objective space only if she is able to encounter equipment in regional-ized places. In addition, people are almost always dealing with region-alized placed equipment (e.g., the clothes they wear) even when they are staring at, observing, or theorizing about occurrent things. Yet, conversely, although Heidegger did not acknowledge this, objective space, too, is something with which people omnipresently deal. As a person acts, she must negotiate an objective space that embraces the present-at-hand things that equipment always already are. This fact also shows that when humans cope with objective space they need not be aware of it—pace Heidegger, they need not be in the cognitive mode of being-in. What all this implies is that Heidegger did not sufficiently think through the relations between lived space and objec-tive space or between regionalized placed equipment and objectively spatial occurrent things. Although Heidegger is to be credited for showing that the space of being-in-the-world is real and central to human life, he bequeathed to spatial theory the task of grasping the relations between this space and its objective counterpart.

On one point, however, Heidegger was emphatic: present-at-hand (occurrent) things are not the foundation of the world. Descartes's analysis of the world is the paradigmatic position Heidegger contested on this point. Descartes conceived reality as composed of three sorts of substance: infinite substance (God), conscious substance (consciousness), and extended substance (things). Setting aside God for present purposes, it follows, according to this conception, that the being of all entities other than consciousness is extension. To be is to be extended, and all further properties of things—for example, motion—are modifications of extension. Extension, moreover, defines space; all subspaces and spatial properties pertain to the domain of extended substances, which Descartes dubbed "nature." Since conscious substance lacks extension, it is aspatial. The famous opposition between subject and object is thus, among other things, an opposition between mind/spirit (*Geist*) and space.

Descartes treated nature as simply the sum-total of extended substances—and nothing more. Nature does not boast a formative structure such as the whole of ends, projects, and uses that defines the being of, and thereby organizes, the equipment that belongs to a world. Instead, each extended substance is what it is by virtue of itself, that is, its extension. It follows that any entity other than consciousness is grounded in a part of nature that serves as the substratum that bears the properties defining it. In particular, since equipment are not of consciousness, they, too, must be analyzed as a certain type of extended object: equipment are material substances to which value (*Wert*) predicates are assigned (SZ 98-9).

Heidegger found much to object to in this metaphysical picture: Descartes' unreflective appropriation of the notion of substance from Medieval philosophy, his failure to consider the being of the substances he postulated, his stark opposition of conscious and extended substances (subject and object), and his attempt to ground all entities other than consciousness in extended substances. Heidegger, however, did not deny that extended entities exist (they are a species of present-at-hand entity), nor that objective space exists, nor that extended entities and objective spaces are pertinent to human life. He simply pointed out that the world wherein people, firstly and mostly, live is composed of near and far equipment in regionalized places. He also held that equipment and their spatiality are not grounded in extended things and their spaces. As suggested, presence-at-hand and arrays of points cannot explain or make intelligible readiness-to-hand and regionalized places. Nor can occurrent things or their relations explain equipment or the interrelatedness of equipment and places. The world wherein people firstly and mostly proceed is founded, not on present-at-hand things, but instead on ends, projects, and uses. Consequently, unlike extended (present-at-hand) things and their objective spaces, the world and its spatiality are not independent of people. As noted, however, securing the nonreductive distinctiveness of the practical world from nature leaves open the task of thinking through the relationship between them.

Before leaving *Being and Time*, I want briefly to discuss the social dimension of the spatiality of being-in-the-world. Heidegger did not directly write anything on this topic, but his remarks about sociality have important implications for it.

Heidegger's analysis of human existence in *Being and Time* is monadic: it is an analysis of the life, or existence, of an individual (functional) human being. His analysis of the spatiality of being-in-the-world is likewise monadic, an analysis of the spatiality of individual existence. This fact does not entail, however, that existence and spatiality are not social in character.

Heidegger contended that equally primordial (*gleichursprünglich*) with being-in-the-world as a constitutive structure of existence is being-with (*Mitsein*). Being-with is coexistence, which in English can also be expressed with the term "sociality" (on at least some uses of it). To say that being-in-the-world and coexistence are equally primordial structures of existence is to say that individual human existence, the object of analysis in *Being and Time*, is essentially in-the-world with others—it is essentially social. I should point out that coexistence is not the same as interaction or bodily copresence. People can coexist even if they are not interacting or perceptually present to one another. Couples, for instance, coexist when one is way on a trip, just as teammates coexist during the off-season.

Being-with has four basic aspects: (1) encountering others within the world, (2) acting toward others, (3) acting in the same world(s) in which others act, and (4) the sameness of the structure of a world for whomever is in it.

The facts that people come across and act toward one another are obvious and require no commentary. To say, meanwhile, that someone acts in a world in which others act is to say that a world that the first person is in is the same world that others are in. The sameness of this world can be treated as a matter of different people encountering and having to do with one and the same entities. When two mechanics work on a car, for instance, the car the one works on is the same car the other works on. This world, in short, is a common, public world. To say, finally, that the structure of a common, public world is the same for whomever is-in-it is to say that the same one complex of ends, projects, and uses governs the activities of anyone who is-in-that-world. This complex delimits, among other things, the uses people make there of equipment and the regionalized places of this equipment. People not only use the same equipment, but also use this equipment in the same range of ways, places, and regions.

The range of uses, places, and regions concerned is defined by *das Man*. In the previous chapter, I described *das Man* as the normativity that governs the life of a particular group of people ("one does this," "one thinks that"). This normativity specifies not just what people should do, but also what it is acceptable for them to do. *Das Man* denotes a range of ends, projects, and uses that are acceptable to or enjoined of a group of people, thus a range of acceptable or enjoined regionalized places for equipment. It follows that to say that the structure of a common, public world is the same for whomever is in it is to say (1) that anyone in that world proceeds attuned to and (2) that the world involved is dependent on the same one *das Man* field of possibilities.

The implications of this analysis for the sociality of spatiality are straightforward. Each person in a given world orientedly encounters near and far regionally placed equipment. As discussed, moreover, there is a certain leeway (*Spielraum*) for the nearness and farness of the equipment a person encounters to change. Both this leeway and the range of places and regions in which equipment can be encountered are delimited the same for whomever is-in-the-world by the normative *das Man* structure of that world. In other words, a single range of possible nears, fars, places, and regions governs what they do: anyone in-that-world proceeds amid one and the same entities in one and the same range of enjoined or acceptable near/far and regionalized places, her specific orientations and the differential near and far of particular entities vis-à-vis her activities depending on what she is currently up to. Heidegger's account of being-in-the-world is an analysis of individual existence, but because existing is as much being with others as it is being-in-the-world, lives proceed, firstly and mostly, amid the same entities within the same field of possible orientations, dis-tances, and regional places.

The Place and Site of Being

Spatial concepts are important in the post-*Being and Time* phase of Heidegger's work. Indeed, two of the most prominent concepts of this phase, place (*Ort*) and site (*Stätte*), are spatial in character. Unlike *Being and Time*, however, which devotes a chapter to space and spatiality, Heidegger's later work rarely pauses to analyze these concepts. Heidegger tends simply to invoke them without explanation, even as what plays the role of place or site metamorphoses. Moreover, the analysis of lived space in *Being and Time* does not resurface in the later work. Some of the basic features of the spatiality of being-in-the-world reappear as key features of place and site, but they do so with transformed meanings. They are no longer features of lived space understood as the space of human experiential acting. Place and site are, instead, associated with lived spaces of the fourth type distinguished in the opening section: spaces with which human life is involved, in terms of which it proceeds. In particular, Heidegger described the lived spaces associated with places and sites as spaces of human dwelling. The features of lived space analyzed in *Being and Time* that reappear in his later work thus do so as features of dwelling spaces.

The basic contours of Heidegger's later analyses of space are contained in the fundamental ontological change that marks the transition from phases two to three of his thought. As described in the previous chapter, in phase two Heidegger treated the clearing of being as the same as Dasein's understanding of being, or its temporality. The light of Dasein's understanding illuminates (*erleuchten*), lights (*lichten*), the clearing. In clearing the clearing, Dasein's understanding is the same as it. In phase three, the clearing and Dasein's understanding of being are no longer the same. People are now construed as *standing into* the clearing, whose happening is distinct from their standing into

it. The clearing just happens: there is being (*es gibt Sein*). The clearing, however, cannot happen without people standing into it—the
clearing befalls people. Indeed, something connected to people is the
place (*Ort*) where it happens.

This change undergirds a transformation in basic spatial sensibility. In phase two, spatiality is at once the spatiality of being-in and the
spatiality of the world; to distinguish between the two is only to emphasize different aspects of a single spatiality, that of being-in-the-
world. In phase three, spatiality (*Räumlichkeit*) is no longer the chief
spatial concept. As mentioned, something connected to human life is
now construed as the *place (Ort)* where the clearing happens. Much of
later Heidegger's thought can be interpreted as analysis of this place.
The switch from spatiality to place does not just signal a decentering
of the clearing away from people. It also implies a transformation in
what space is all about, from expanse to place: from that *wherein*
something of interest occurs to that *where* it does so. What's more,
that which plays the role of the place at which the clearing happens is
no longer, as in *Being and Time*, an aspect of people's being: understanding or temporality. What plays this role are, instead, entities
such as works of art, built structures, and poetry in a broad sense. In
phase three, finally, two types of space bound up with place and site
are analytically distinguished. The first type embraces the spaces of
human dwelling. The second type encompasses the spaces that—
together with time—constitute the opens that make up clearings of
being. Spaces of the both sorts are intimately related to the sites where
the clearing happens; in the end, the two types of space are one.

An animating ontological intuition in Heidegger's later philosophy
is that something connected to people is the place where the clearing
happens. The clearing happens, in other words, at something in the
clearing—a particular entity. In serving as the place of the happening
of the clearing, the entity in question is said to found or institute
(*stiftet*) the clearing. At first Heidegger understood this idea in the
terms of *Being and Time*: the place where the clearing occurs is
Dasein, especially its understanding. This way of thinking persisted
into the 1930s. Sometime early in that decade, however, Heidegger
began to think that the entities concerned are not human beings. The
first decisive presentation of this intuition is found in a famous 1935–
36 essay, "The Origin of the Work of Art," which analyzes artworks as
entities at which the clearing happens. In the 1930's and 40s, mean-

while, Heidegger also deemed the polis to be such an entity and began to acknowledge entities in the built or natural environment—for example, rivers and bridges—as such. Thereafter, in the 1950s, "things" (*Dinge*) became the general name for the entities involved: as discussed in the previous chapter, when things thing (*Dinge dingen*) the clearing is instituted (in Hölderlinian language, the fourfold are unified into a mirror-ring play). The thinging thing is the place at which the clearing happens. Throughout these changes, finally, and beginning at least with the essay on artworks, poetry and language are the ultimate entities at which the clearing happens. Indeed, artworks, rivers, bridges, and language play this role only insofar as, in their essence, they are poetry (Heidegger dropped references to the polis after the early 1940s).

The nonhuman entities at which the clearing happens, that make up the place of being, are called "sites" (*Stätten*). As just noted, the idea that entities other than people are sites made its first appearance in Heidegger's account of artworks. "Art," he memorably wrote, "is truth setting itself into the work."[m] (OWA 39, translation slightly corrected). I discussed in the previous chapter Heidegger's idea that the truth of beings is their unconcealedness (*Unverborgenheit*). Art institutes unconcealedness by way of unconcealedness setting itself into the work. Heidegger conceptualized this achievement as the setting up (*Aufstellen*) of a world, where a world is a "space" for human dwelling. Setting up a world via an artwork also requires setting forth (*Herstellen*) an earth (*Erde*), the material stuff of the artwork, which holds itself back in favor of the world. Upon this earth, and as the basic structure of this world, are gathered and joined (*gefügt*) those "paths and relations in which birth and death, disaster and blessing, victory and disgrace, endurance and decline acquire the shape of destiny for human being."[n] (OWA 42) An artwork sets up world for human dwelling because setting forth the work—a particular formed material entity—gathers these basic paths and relations. In this world, all things receive their "lingering and hastening, their remoteness and nearness, their scope and limits"[o] (OWA 45).

I mentioned in the previous chapter that Heidegger, in his third phase, held that truth occurs as clearing concealing (*lichtend Verbergen*), as a mix of uncovering and covering up. His analysis of the work of art well illustrates this conception of truth. For the joint setting up of the world/setting forth of the earth accomplished in the work of art is, precisely, the institution of an open (world) via a concealing (earth): "Setting up a world and setting forth the earth, the work is the contention of the strife [between clearing and concealing-TRS] in which the unconcealedness of being as a whole, or truth, is won."[p] (OWA 55, translation modified)

In this essay, Heidegger contended that truth happens only in such a way that it establishes itself in (*sich einrichtet in*) the open that is won in the strife of clearing concealing. In order for truth to happen, in other words, there must be an entity in the open at which truth "takes its stand and attains its constancy."[q] (OWA 61) Indeed, the opening of the open and the self-establishing of truth in an entity in the open are inseparable. "The establishing of truth in the work is the bringing forth of a being... The bringing forth places this being in the Open in such a way that what is to be brought forth first clears the openness of the open into which it comes forth."[r] (OWA 62). Artworks, however, are not the only entities in which truth establishes itself. Others are the deed that founds a state (*staatgründende Tat*), the essential sacrifice, and the questioning of the thinker. Any entity, furthermore, that accomplishes this is capable of doing so because it is in essence poetry.

The connection between the clearing and poetry/language is crucial for Heidegger. "As sensuous meaning, the word traverses the expanse of the leeway between earth and sky. Language holds open the region in which humans, upon the earth and beneath the sky, dwell in the house of the world."[2,s] (translation slightly modified) Ultimately, all entities at which the clearing happens are poetry, even the rivers at which this occurs, such as the Danube or the Rhein. Entities that institute a clearing qualify as poetry because the essence of poetry is to take the measure, to name the holy, to articulate what is sent (*Zugeschickte*), which measures out, or opens up, the between between earth and sky where humans dwell. Precious little of what is conventionally dubbed poetry qualifies as genuine poeticizing. I shall basically put aside this aspect of Heidegger's thought in the following.

Heidegger's acknowledgement of deeds that found states as entities into which truth sets itself reflects his idea, in the 1930s and 40s, that

---

2 Martin Heidegger, "Hebel—Friend of the House," trans. Bruce V. Foltz and Michael Heim, in *Contemporary German Philosophy*, Volume 3, Darrel E. Christenson et al. (ed), University Park, The Pennsylvania State University Press, 1983, pp. 89-101, here p. 101; *Hebel, Der Hausfreund*, Pfullingen, Neske, 1957, p. 38.

the polis is a site of being. Indeed, some formulations at this time describe the polis as *the* site of being. Heidegger could claim this because he rejected the standard rendering of "polis" as polity, state, or city-state. The polis, for him, is a broader entity, embracing everything about human life: "the gods and the temples, the festivals and the games, the rulers and the council of elders, the people's assembly and the armed forces, the ships and the field marshals, the poets and the thinkers"[t] (Ister 82). Everything subject to the basic "paths and relations" in which a historical people wins the shape of its destiny belongs to it. Unlike Heidegger's discussion of artworks, moreover, his remarks on the polis clearly reveal the entanglement of humanity with the happening of the clearing. The polis, he wrote, is the site of being: "the in itself gathered site of the unconcealedness of being as a whole,"[3,u] in which "all entities and all comportment toward entities gathers itself"[v] (Ister 86, translation modified). The polis is also, at once, the site of the historical sojourn (*Aufenthalt*) of humans in the midst of entities. It is "the site, the there, in which and as which Dasein is as historical. The πόλις is the site of history, the there, *in* which, *out* of which, and *for* which history happens."[4,w] In particular, the polis is the abode (*Ortschaft*) for the historical sojourn of Greek humanity.[5] These quotations reveal that the entities at which the clearing opens are at once entities amid and with (*bei*) which humans carry on. They are entities at which the clearing happens and humans stand into it, where humans receive the sending (*Schickung*) into which they are fitted: "The πόλις is the site that belongs to historical humanity, the where, to which humanity as ζῷον λόγον ἔχον belongs, the where, from which alone the structure is inflicted upon him in which he is placed."[6,x] The clearing happens only in befalling, in being received by, a group of people.

A couple of notes before continuing. First, in the 1930s Heidegger described the people who stand into the happening of the clearing as a

---

3   Martin Heidegger, *Parmenides*, trans. André Schuwer and Richard Rojcewicz, Bloomington, Indiana University Press, 1993, p. 89-90; *Parmenides*, Gesamtausgabe 54, Frankfurt am Main, Vittorio Klostermann, 1982, p. 133.
4   Martin Heidegger, *An Introduction to Metaphysics*, trans. Ralph Manheim, New Haven, Yale University Press, 1959, p. 152, translation modified; *Einführung in die Metaphysik*, Tübingen, Max Niemeyer, 1966, p. 117.
5   Martin Heidegger, *Parmenides, op. cit.*, p. 90 (German p. 133).
6   *Ibid.*, p. 95 (German p. 141).

people (*Volk*). By the early 1940s, he ceased referring to these people as peoples (*Völker*) and left it unspecified (e.g., *Menschentum*) what sorts of group might be involved. This development continued in the 1950s, when Heidegger appropriated Hölderlin's language to denote the people involved simply as mortals (*Sterblichen*). Second, however, Heidegger portrayed the life of the people to whom a clearing occurs as extremely unified. This portrayal perpetuated the idea of *Being and Time* that the clearing (and spatiality) in which people firstly and mostly carry on is a common normative clearing. To be sure, soon after *Being and Time* Heidegger ceased analyzing the existence of individual people and theorized more collectively. An example of his more collective thinking is the idea that groups of people, not individuals, stand into the clearing. But although he no longer needed such maneuvers as complementing being-in-the-world with being-with and building sameness and commonality into being-with in order to secure the social character of the clearing, the clearing remained unified. In the above quotation from *Parmenides* 95, this unity is expressed in the notion of *a* structure (*Fug*), *a* historical sending, in which all the various phenomena of culture and society are gathered. Heidegger, in short, never disabused himself of the idea that it is appropriate to analyze human socioculture as unified ensembles. In fact, because the people who stand into the happening of the clearing must share a language, and at least many customs and probably a history too, it is not clear that Heidegger ever abandoned the idea that the groups involved are peoples (*Völker*).

A particularly clear discussion of abode as the space of a people is found in Heidegger's lectures on Hölderlin's poem, "The Ister." In these lectures, Heidegger claimed that it is essential to the history of a people that that people begin distant from itself and then return to itself via a passage through what is foreign to it. When a people has returned to itself through such a passage, it is at home (*heimischsein*). Heidegger maintained that the being at home of a historical humanity involves a certain sort of space, more precisely, a certain sort of time-space. This time-space is the time-space in which the people dwells. This time-space is also at once the time-space that constitutes the open of the clearing into which this people stands. To explicate this time-space, Heidegger conceptualized place (*Ort*) as the here of human dwelling and abode (*Ortschaft*) as the way a place is a here, that is, as the happening that belongs to that place as the here it is. (Much later,

Heidegger defined abode as the "interaction [*Zusammenspiel*] of places").[7] Toward the end of these lectures, Heidegger stated that abode is the place that belongs to being at home, i.e., home (*Heimat*). Time, meanwhile, is conceptualized via the notions of journey (*Wanderung*) and travels (*Wanderschaft*). A journey is, simply, movement into past and future: it is the movement of time that joins past and future. The life of a historical people is the prime example in the Ister lectures (as individual human life was in *Being and Time*). Travels, moreover, is the fulfilled happening that belongs (*erfüllte Wesen*) to a journey, through which a people comes home. Abode and travels are thus the space and time of the home of a historical people, the dwelling time and dwelling space of that home, thus the time-space destination of that people's historical trajectory. As indicated, abode and travels are also the time and space that co-constitute the open of the clearing into which that people stands when it is home. Note that the clearing is conceived of as a temporalspatial affair. Heidegger thought this throughout phase three of his career, a fact from which I have been abstracting until now. In the Ister lectures this idea is expressed in such claims as that the between (between humans and gods) of a historical people is opened as the abode of its travels and as the travels of its abode.

In the previous chapter, I pointed out that different groups of people stand into different clearings. In the Ister lectures, this idea becomes the thesis that the abode and travels that characterize a historical people are unique to it: the collective life, in the form of which a people comes to itself and is at home, is always a particular accomplishment with its own having-beenness and coming toward (temporality) and its own dwelling spaces tied to the entities amid which it transpires. The historical particularity of these time-spaces greatly contrasts to the assumed universality of the abstract four-dimensional space-time theorized in modern physics. Heidegger linked the emergence of theories of this physical space-time to the onset of the era of modern technology. He also contrasted settlement (*Siedlung*) as an attachment (*Bindung*) to place (*Platz*) with the "space grasping and time gathering proceeding" (*raumgreifenden und zeitraffenden Vorge-*

---

[7]   Martin Heidegger, "Art and Space," trans. Charles H. Seibert, *Man and World*
       3 (1973): 3-10, here p. 6; *Die Kunst und der Raum*, St. Gallon, Erker, 1969, p.
       10.

hen) that is characteristic of the human activity, definitive of that era, that seeks to calculate and conquer the world (Ister 48).

The idea of humanity coming to itself underwrites a notion of near and far different from the nearness and farness of equipment in practical activity. To come to or remain away from itself is for a humanity to be near to or far from itself. These states of affair coincide with the gods being near to or far from that humanity, all these conditions also converging with the nearness or farness of being from humanity. In this context, Heidegger also spoke of the gods or being nearing (*nähren*) humanity and of humanity being near (*in der Nähe von*) being. Further in this vein, he wrote that when humans genuinely dwell, the four components of the fourfold (*Geviert*)—earth, sky, mortals and gods (see previous chapter and below)—near one another as unified. In sum, when things are right in human life, people, dwelling, being, gods, earth, sky, and the unity of these come together, are near to one another. When this happens, humans have come to themselves; they are at home.

At some point Heidegger began to think of entities in the built or natural environment as sites of being. This development holds special significance for the present discussion because, more explicitly than in the case of other sites, Heidegger treated these entities, not just as sites where being happens, but also and at one with this as sites of dwelling: sites of being are at once the entities that admit, or more suggestively, make room for (*einräumen*) human dwelling. Indeed, the essence of the human sojourn on the earth is dwelling, and dwelling involves receiving the clearing. It is when humans dwell with-amid these entities that the clearing happens at them.

The idea that sites have this twofold role, as well as the spatial character of the place (*Ort*) of being more generally, are made concrete in the well-known essay "Building Dwelling Thinking." This essay contains the most focused discussion of spatiality in the third phase of Heidegger's philosophy. In particular, this essay clarifies the character of the spaces that are set out in the here that is set up by a site of dwelling-being. This essay, which also brings together most of Heidegger's Hölderlinian conceptual armature, begins by clarifying the notion of dwelling. By virtue of dwelling, Heidegger wrote, humans, now called mortals, stand into the ring-mirror play of the fourfold. This ring-play occurs when things gather the fourfold (when things thing), and in this essay Heidegger characterized the role of mortals in the ring-dance as safeguarding and preserving (*schonen*) the ring-dance in the happening that belongs to it (its *Wesen*). Doing this involves letting things be (what they are). What letting things be concretely means is twofold: (1) looking after (*pflegen*) the things that grow and also erecting (*errichten*) the things that do not grow—thus, Heidegger claimed, building (*bauen*), whose main two forms are looking after and producing (*herstellen*)—and (2) dwelling amid-with (*bei*) things.

A thing, consequently, is the place (*Ort*) where a clearing opens: it "allow[s] [*verstattet*] a site for the fourfold." (BDT 154) At the same time, a thing is a place of human habitation: a site institutes places and paths that admit, or make room for (*einräumen*), human habitation. These places and paths constitute a here. Heidegger gave a bridge as an example. Along the river, before the bridge was built, there were only many points (*Stellen*) where something could occur or stand. Once the bridge had been built, one of these points became an place— by virtue of the bridge. Indeed, the bridge is this place. (A thing both is and is at a place, just as, in *Being and Time*, a piece of equipment both is and occupies a place [*Platz*].) As a place, the bridge institutes a here, it sets out a variety of places and paths, for example, a way into town, a place to fish, a way into the fields, a place to exchange goods, a way across the river, and the opposite bank of the river. These places and paths, which are differentially near to and far from the bridge, make room for human habitation. In sum, the bridge is a dwelling site (*Wohnstätte*) because it institutes a here that contains an array of near and far places and paths for human activity. These are spaces people go through (*durchgehen*) daily.

In its role as place (*Ort*), a thing admits, makes room for, spaces, for nexuses of places and paths at and along which people act daily. In this regard, things are functionally equivalent to understanding (*Verstehen*) in *Being and Time*—for in that book understanding is credited with admitting, making space for (*einräumen*), regions: that is, for totalities of possible places (*Plätzen*) for equipment that are differentially near or far vis-à-vis human activities. As place (*Ort*), however, a thing is also where a clearing "takes a stand:" the thing admits, makes room for, the fourfold. And it accomplishes this *by way of* setting out places and paths for human dwelling and setting itself up in those spaces (BDT 158). In other words, the space that a thing sets up is at once a space where humans dwell and the space of a clearing. Since, moreover, a here of places and paths cannot be independent of the people who live through it, it follows that the happening of a clearing is not independent of people standing into it. Coordinately, people cannot be without standing into a clearing by dwelling amid things, regardless of whether their activity is commensurate to or diverges from the role allotted to humans in the mirror-play. Standing into a clearing is the happening that belongs to human being.

The dwelling spaces opened about things strongly resemble the regions of activity places described in *Being and Time*. In each case, human life carries on in some open whole or region that is filled with differentially near and far places and paths. In other late essays, for instance, "Conversations on a Country Path about Thinking" ("Zur Erörterung der Gelassenheit"), in *Discourse on Thinking* (*Gelassenheit*), and Heidegger's last lecture, "Art and Space," Heidegger even marshaled the word he had used in *Being and Time* to designate this open realm: "region" (*Gegend*, or *Gegnet*). Throughout his career, consequently, Heidegger treated human life as transpiring amid regionalized places and paths. Corresponding to the ontological shift marking the transition from phases two to three, a significant difference in the two versions of this basic spatial intuition concerns the center around which the region opens. Whereas in *Being and Time* the center is practical dealings, in Heidegger's later philosophy it is things: regions, the arounds of the world-around, the whereins or wheres where human existence proceeds, center on things, not concernful dealings. Humans go through regions, in the sense of acting within and being attuned to them, but the regions are arranged around things, not about their activity. Of course, as just noted, the regions instituted around things, like the regionalized activity places of *Being and Time*, are not independent of human activity: it is by virtue of what people do that such and such places and paths—and not others—are set out about things. But things are the poles around which they are set up. Another important difference between the accounts of regions found in phases two and three is that, after *Being and Time*, Heidegger dropped the assumption that the places and paths that fill regions are defined by teleology/instrumentality alone and not by any other kind of significance, for example, aesthetic or religious.

A further noteworthy parallel with the analysis in *Being and Time* concerns the distinctiveness of objective space from the space of human being. As described, *Being and Time* argues that when things are encountered in objective space, in an objective manifold of points, dimensions, and measurable distances (*Abstände*), entities are shorn of all equipmentality and regional placement. "Building Dwelling Thinking" describes the distinctiveness of dwelling space from objective spaces similarly. People, Heidegger wrote in that essay, not only act in places that are differentially near to and far from such things as bridges, but also treat these places as points between which crossable distances (*durchmessbare Abstände*) occur. When they do this, the nearness/farness of places falls away and is replaced by "distances of interval-space" (*Abständen des Zwischenraums.*) From space in the sense of interval-space, moreover, the familiar manifold of three-dimensional Euclidean space can be abstracted, just as from this three-dimensional extensional space n-dimensional manifolds, pure mathematical constructions, can be abstracted. Throughout his career, Heidegger maintained that objective or abstract spaces cannot account for the spaces of existence or dwelling. It is true that, when encountered, objective spaces (but not abstract spaces, which cannot be encountered) show themselves as having already been there. It is because of this that equipment and places can with truth be described as separated by numerical distances. But this does not mean that objective space grounds the spaces of existence or of dwelling or that objective space is the true reality either upon which human spaces are projected or out of whose properties the characteristics of human spaces are constructed (as in Descartes). The two categories of space are distinct. Indeed, Heidegger maintained that dwelling space, like the space of being-in-the-world, retains a twofold priority over objective space: (1) it is the space people are automatically in as people and (2) whereas dwelling space can be encountered as an objective space (in the cognitive mode of being-in), objective space can never be encountered as dwelling space. That is, whereas dwelling space contains objective space in potentia, objective space eliminates all traces of dwelling space.

According to Heidegger, moreover, the increasing tendency in the contemporary world to treat space exclusively as a three-dimensional extensional manifold is part of the attempt in the age of modern technology to master and set everything up as standing reserve (*Bestand*) for the satisfaction of human desire and need (ultimately, for the sake of ordering per se). Central to this process of mastering and organizing is the mathematical-technological idea that any part of lifeless nature is ordered in the sense of being calculably aligned (*berechenbar zugeordnet*) with every other part (see Ister 39-42). This idea treats space as a coordinate-system with which the position of any element can be specified and the alignment of this element to anything else calculated. As science and this conception of reality increasingly dominate Western people's understanding of reality, this construal of space becomes primary. The very existence, let alone primacy, of lived and dwelling space becomes forgotten. These spaces thus suffer the fate of the clearing of being at the culmination of metaphysics. Although the development of modern technology has overcome distances, has brought close much that was far, and is tending toward making everything equally near and far, genuine nearness—sensitivity to and care for dwelling space in conjunction with openness to being and the gods—remains absent.[8]

The Open as Space

Heidegger never directly discussed the following idea, but it seems obvious that what is arguably his second master concept, that of the clearing (*Lichtung*) of being, is a spatial concept of some sort. Indeed, it resembles the concept of absolute objective space. It differs from the latter because the clearing, unlike absolute space, is not an entity; it is not a container that holds positions or entities. Nonetheless, it is an open realm in which entities show up. It is *where* beings are. As mentioned in the previous chapter, Heidegger wrestled with the issue of how to talk about the clearing of being without thereby hypostasizing it as an entity. His most prominent solution to this problem was to say of the clearing, not that it is, but that it happens: the clearing is an ontological space that opens via clearing concealing (*lichtend Verbergen*). The following chapter will discuss the wide appropriation that this peculiar type of space has enjoyed in subsequent philosophical and social thought.

Near the beginning of the previous section, I wrote that a second type of space emerges in phase three of Heidegger's career: space as a constituent structure of the clearing. I have already alluded to this

---

[8]   See Martin Heidegger, "The Thing," in *Poetry, Language, Thought*, trans. Albert Hofstadter, New York, Harper, 1971, pp. 163-86, here pp. 163-4; "Das Ding," in *Vorträge und Aufsätze*, Pfullingen, Neske, 1954, pp. 157-80.

type of space in characterizing the space set up about a thing as the space of a clearing. Heidegger wrote relatively little on this topic. In "Building Dwelling Thinking," for instance, he said no more than that the space set up about a thing admits the fourfold. Heidegger's lengthiest and clearest discussion of this topic occurs in the *Contributions to Philosophy* of 1936–38, which some scholars today consider to be Heidegger's second magnum opus.

A basic idea of Heidegger's thought in its third phase is that truth is always accompanied by untruth, that unconcealment (*Unverborgenheit*) always involves concealment (*Verborgenheit*). He claimed, for instance, that the unconcealedness of an entity involves the simultaneous concealedness of being. As a result, Heidegger—as noted—often described the clearing (*Lichtung*) as opening via clearing concealing (*lichtend Verbergen*): beings are, in the clearing/lighting of (i.e., that is effected by) the self-concealment of being. In the *Contributions to Philosophy*, Heidegger claimed that the open that constitutes the clearing is time-space: time-space is the "light of the concealing withdrawal " (*Licht der verbergenden Entzugs*) in whose illumination whatever is, whatever is unconcealed, shows up. All entities, consequently, are temporal-spatial in nature. Time-space, Heidegger further claimed, is the "first clearing of the open as the empty"[y] (C 265). This empty (*Leere*) is neither an unoccupied void nor the unoccupiedness of—the lack of occurrent entities within—"the frames and forms of order for the calculable presence of space and time" in the age of modern technology. This empty is, instead, the "original gaping open" (*ursprünglich Aufklaffung*) of the between between humans and the gods, i.e., the clearing. Among other characteristics, this open is tuned attuning (*gestimmt stimmend*) and joined joining (*gefügt fügend*), meaning that it embraces basic orientations and dispositions according to which the humans standing into it are oriented and disposed. The empty is spatial, moreover, because it is a capturing (*berückend*) surrounding hold (*Umhalt*) through which the humans standing into it are set firmly in place. The empty is a surrounding hold by way of being rooted in particular entities (cf. the work of art essay), and for this reason Heidegger described the spatial component of the clearing as a site (*Stätte*). This empty is temporal, meanwhile, because it moves into futurizing (*entrückend in Künftigkeit*) and is "thus at once breaking open something that has been" (*damit zugleich aufbrechend ein Gewesendes*), that is, it involves both coming-toward and having-beenness, awaiting and remembering (C 268, translation altered). Futurizing and having-beenness, finally, also meet up and constitute the present, which Heidegger described with the term "moment" (*Augenblick*). The time-space of the empty, of the open that constitutes the clearing, is thus moment-site (*Augenblicks-Stätte*).

Time-space is the empty, the open, of the clearing of being. It is a basic feature of the clearing. It is obvious that the time-space so conceptualized is not objective time-space, for example, the four dimensional coordinate system. That this time-space is tuned, joined, futural, having-been, of the moment, and a surrounding hold indicates that it pertains essentially to humanity: it is also the time-space of human being. Time-space, accordingly, is at once a basic structure of the clearing and a basic feature of human being. In the *Contributions to Philosophy*, however, Heidegger was not clear about just how time-space is a human phenomenon. He wrote almost nothing about sites, and the notion of a moment seems more to do with the expanse of history than with ongoing human life. Regarding sites in particular, it is not clear whether people dwell in sites in the way that the essay "Building Dwelling Thinking" describes this—sojourning at things and going through the spaces things institute. Compounding this interpretive uncertainty is a lack of clarity about whether Heidegger, in the *Contributions to Philosophy*, thought that the moment-site, the clearing, and the standing of humans into the clearing are features of the universal condition of humanity or, instead, features of a possible future situation that will come about only when humans act commensurately with the role allotted to them in the mirror-play. "Time-space is the capturing-moving-toward gathering surrounding hold …whose essential happening becomes historical in the grounding of the 'there' through Da-sein (its essential ways of sheltering truth [in entities— TRS]"² (C 269-70, translation altered). Statements such as this are ambiguous between these two options. It is not clear whether the light that clears the clearing in its happening opens the expanses of the history and dwelling realm of a people or shines only at the end of that people's history.

No where else did Heidegger focus on time-space in this way. He interjected the concept into a late lecture, "Time and Being" (1962), which is found in the book *On Time and Being*. The time-space he discussed, however, is the span of the mutual reaching, or "nearing nearness" (*nährende Nähe*), of the three dimensions of time—the expanse, or breadth, of time, thus a feature of time. Indeed, this essay assigns a fundamental role to time in the happening of the clearing that parallels the fundamental role that *Being and Time* assigns to time vis-à-vis being: whereas the earlier work makes time the meaning of being, the later essay anoints time the "prespatial abode" (*vorräumliche Ortschaft*) of the clearing of being (TB 14).[9] Apart from the intriguing spatial characterization of the unity of time as nearing nearness (which parallels the description of the components of the fourfold as nearing one another when things thing), Heidegger left unmentioned whatever fundamental role space might play in the clearing. Toward the end of the lecture, he simply noted the need for an analysis of space that relates it to the event (*Ereignis*), referring the reader to the essay "Building Dwelling Thinking." In the *Contributions to Philosophy*, by contrast, space is distinct from, and not the space of, time. In that book, Heidegger clearly distinguished temporalizing (*Zeitigung*) from spatializing (*Räumung*); time-space, he wrote, is "the in itself temporalizing-spatializing-oscillating moment-site of the 'between,' as which Da-sein must be grounded."[aa] (C 271, translation altered) Given what little Heidegger wrote on this topic, the analysis in "Building Dwelling Thinking" can be taken as filling in the very abstract characterization of space found in the *Contributions to Philosophy*.

All in all, three basic characterizations of space are analytically distinguishable in Heidegger's analyses of lived space: (1) space as the aroundness of the world in which people proceed, composed of the regional places of entities differentially near and far vis-à-vis people's activity; (2) space as the place (*Ort*) where the clearing happens, embracing a here of human dwelling that is centered around things and contains places and paths for dwelling; and (3) space as a component of the empty, or open, of the clearing in which entities are. In the shift from phases two to three of his thought, the second sort of space supplants the first as the one most pertinent to the question of being. It and the third sort of space are thereby profoundly entangled. One might argue that the first sort of space coincides with the third in *Be-*

---

9   Martin Heidegger, "Time and Being," in *On Time and Being*, trans. Joan
      Stambaugh, New York, Harper, 1972, pp. 1-24, here 16; "Zeit und Sein," in
      Zur Sache des Denkens, Tübingen, Max Niemeyer, 1969, pp. 1-26, p. 16.

ing and Time, but this speculative argument would take us beyond the tasks of the present interpretation.

5 Legacies

Heidegger's ideas cast long shadows over twentieth century intellectual history. He is generally considered to be one of the two philosophers most responsible for the direction of philosophical work in that century (the other is Ludwig Wittgenstein). His analysis of human existence as being-in-the-world, for example, played a decisive role in enabling Western philosophy, and Western thought more generally, to overcome a Cartesian conception of human being. His ideas have also reached well beyond philosophy into literary theory, theology, geography, sociology, political theory, cognitive science, architecture, and ecology. Heidegger's thoughts on space have likewise inspired, oriented, and been appropriated by theorists in various disciplines. Although it is too strong to claim that these conceptions enormously influenced these fields, they have spawned fruitful—and continuing—lines of research across them.

It is impossible to gauge the difference Heidegger's Nazi period has made to his reception. Some thinkers and intellectuals have refused to engage his ideas because of his involvement with National Socialism. So tainting, moreover, is the stain of collaboration, of membership in the party, and of Heidegger's silence after the war that some writers who appropriate his ideas, or even just discuss them, feel obliged to provide elaborate defenses, or heartfelt reflections on the propriety, of doing so. Of course, other interpreters and appropriators who denounce Heidegger's behavior so disengage his philosophical views from his political attitudes and activities that they can address his philosophy without feeling the need to defend or excuse doing this. And there are even some commentators who are convinced, conversely, that Heidegger's philosophy is thoroughly fascist but who also believe that philosophical thought cannot but proceed through a confrontation with his ideas. In any event, it is possible that Heidegger's ideas on space would have enjoyed a greater legacy if he had not so stigmatized himself.

In Heidegger's Wake

The importance of Heidegger's thoughts on space for the subsequent history of spatial theory is not nearly as great as is the significance of his philosophy for subsequent thought more generally. It will be useful, consequently, to begin with the latter.

Being and Time is a monumental work. Above all, three major intellectual developments in the 20th-century can be directly traced to it. A fourth development that it instigated is philosophical attention to and invocations of being, but this development is too diffuse and uneven to be addressed in the present work. The first development which I will discuss is philosophical hermeneutics. Hermeneutics, as a general theory of the interpretation of both texts and those who write them, is the child of the 18th- and 19th-century German classicist, theologian, and philosopher Friedrich Schleiermacher. A century later, the German philosopher, psychologist, and historian Wilhelm Dilthey christened hermeneutics the methodology of the historical human sciences, whose task, Dilthey claimed, is to understand and interpret objective expressions of mental-spiritual (*geistige-seelische*) life. In *Being and Time*, by contrast, Heidegger made understanding a—arguably the—principal component of human existence. Unlike his predecessors, Schleiermacher and Dilthey, he did not treat understanding and interpretation (explicit understanding) as two cognitive acts among others, for example, description and explanation. He claimed, instead, that every encountering of something is informed by an understanding of the being of the encountered entity: that is, that anything encountered is encountered *as*, i.e., understood as, such and such. So construed, understanding and interpretation become an omnipresent and constitutive dimension of human life. And hermeneutics, as the doctrine and description of understanding and interpretation, is transformed from the methodology of the historical sciences that it was in Dilthey into a comprehensive account of human existence.

Subsequent theorists writing in or about hermeneutics such as Hans Lipps, Emilio Betti, Paul Ricoeur, David Hoy, and Richard Palmer (see bibliography), have either presupposed Heidegger's so-called "ontologization" of understanding or opposed it as a singularly important, but regrettable development. The thinker who is perhaps most thought of as affirming Heidegger's reconception of understanding is the German philosopher Hans-Georg Gadamer. Gadamer transformed Heidegger's theory of the forestructure of understanding (see chapter three) into a theory of both history and the historical character of human life. Whereas Heidegger had stressed the projective dimension of human existence, Gadamer highlighted its thrown dimension: the fact that a human life is always already such and such. According to him, the prestructure of understanding, a facet of this thrownness, signals the embedment of human life in historical traditions. Traditions are at once sources of the prestructures of understanding and phenomena whose cognitive infrastructure is constituted by the perpetuation of these prestructures. The thesis that all understanding and interpretation are beholden to tradition was not a new idea, but it thereby received a decisive new injection of energy. Because of Gadamer's work, above all his 1960 book *Truth and Method*, most contemporary humanist theorists writing on understanding, interpretation, history, or the philosophy of the human and social sciences affirm—often sotto voce—this Heideggerian interpretation of the conditionality of human understanding and the finitude of human existence. At the same time, considerable controversy surrounds exactly what follows from this.

A second development that can be laid at Heidegger's doorstep is existential psychology. *Being and Time* opposed accounts of human life that centered human existence in a self-transparent realm—called consciousness or mind—that is separate from the external world. Those psychologists and psychoanalysts who were struck by this analysis were thereby robbed of the traditional conceptions of mind with which they had worked. Any psychonaut who read *Being and Time* was also highly unlikely to believe that the concepts and methods of the natural sciences are adequate for grasping the psychological dimension of human existence. For some of these students of the human psyche, the ideas of Edmund Husserl, Jean-Paul Sartre, Martin Buber and, above all, Heidegger provided an alternative. Existential psychology, which was a lively enterprise in the middle of the 20th-century, studied both the normal and the pathological forms taken by human being-toward entities within the world. Particularly concerned with consciousness and neurotic behavior, it studied these phenomena as aspects of being-in-the-world: existential psychological analyses of individual lives encompassed detailed investigations of "the individual's mode of being-in-the-world," the particular thrownesses, fallings, and projections that make up particular lives. The most prominent works of existential psychology were Ludwig Binswanger's *Grundformen und Erkenntnis menschlichen Daseins* (1942) and Medard Boss's *Psychoanalysis and Daseinanalysis* (*Psychoanalyse und Daseinsanalytik*, 1957) and *Existential Foundations of Medicine and Psychology* (*Grundriss der Medizin und der Psychologie*, 1971).

A third and diffuse current of thought spawned by *Being and Time* can be vaguely characterized as analyzing human existence, or understanding human life, as being-in-the-world. A large number of theorists after 1927 appropriated Heidegger's thesis that the fundamental structure of human life or existence is this. This thesis did not so much generate a particular line of thought or succession of thinkers as disseminate as a general understanding underlying much European humanistic theory. The 1945 magnum opus of the prominent French phenomenologist Maurice Merleau-Ponty, *Phenomenology of Perception*, is a fine example. Although in many ways strongly Husserlian in character, this book is informed by the intuition that being-in-the-world forms the overall setting in which its objects of analysis occur (e.g., attention, judgment, the body, sexuality, language, space). Furthermore, although human existence as such has not often been a topic for French "poststructuralists" such as Jacques Derrida, Michel Foucault, Luce Irigaray, Emmanuel Lévinas, Jean-François Lyotard, Jean-Luc Nancy, and Chantal Mouffe, such thinkers have assumed and built upon Heidegger's master intuition in their very different analyses of human life and the contexts in which it transpires. This cannot be said of other French thinkers such as Gilles Deleuze and Michel Serres, who had considerably less affinity with Heidegger's analysis of existence. Still, Deleuze's concept of event bears uncanny resemblance to Heidegger's notion of the event. In the first third of the 20th-century, finally, various prominent Protestant theologians, above all, Rudolf Bultmann and Paul Tillich, appropriated Heidegger's analysis of the structure of existence. Indeed, their brand of theology came to be known as existential theology.

Beyond phenomenology, poststructuralism, and theology, two kinds of "philosophical anthropology" have appropriated Heidegger's analysis of existence. The first kind comprises philosophers who develop a theory of human beings that is philosophical in scope and conceptualization but strongly informed by the latest ideas in anthropology, sociology, and biology. This form of theorizing was prevalent especially in Germany in the middle of the twentieth century; Helmuth Plessner, for example, is a prominent philosophical anthropologist who is ultimately Heideggerian in orientation (cf. *Die Stufen des Organischen und der Mensch* [1928]). Another text in this vein is the American philosopher Marjorie Greene's *Approaches to a Philosophical Biology* (1968). The second kind of philosophical anthropology in question includes the work of contemporary English anthropologists and archaeologists who have developed Heideggerian accounts of human life. Prominent exemplars are Tim Ingold (*The Perception of the Environment: Essays in Livelihood, Dwelling, and Skill*, 2000), Julian Thomas (*Time, Culture, and Identity: An Interpretive Archaeology*, 1996), and Christopher Gosden (*Social Being and Time*, 1994).

A third group of theorists have appropriated Heidegger's account of existence as the basis of analyses of human activity and intentionality. The most prominent such theorist is the American philosopher Hubert Dreyfus (*Being-in-the-World: An Interpretation of Heidegger's Being and Time, Division One*, 1991). Others are the English sociologist Anthony Giddens (*Central Problems in Social Theory*, 1976) and the American philosophers John Haugeland (*Having Thought*, 1998) and Taylor Carman (*Heidegger's Analytic*, 2003). According to Dreyfus's influential interpretation of *Being and Time*, Heidegger developed an account of ongoing human life that highlights the dependence on embodied skills of the largely successful behavior a person engages in in her moment-to-moment existence. On this account, knowledge and cognition play a distinctly secondary, even somewhat remedial, role in the production of everyday life. Dreyfus's championing of skills over cognition directly challenges a variety of research programs in cognitive science and artificial intelligence. Dreyfus's interpretation of *Being and Time* also draws Heidegger close to his contemporary Wittgenstein (as well as to other defenders of abilities over knowledge such as the English philosopher Gilbert Ryle and the American philosopher and economist Michael Polyani).

This short survey suggests how *Being and Time* has shaped theory in the humanities and humanist social sciences. The main ideas of his philosophy after *Being and Time* have also been influential, though the disciplines they have molded cluster more narrowly in the humanities. A prime example of a later idea of Heidegger's that has enjoyed a long afterlife is his thesis that language is the site of being ("Language is the house of being"). The philosophies of Gadamer and of Derrida are two prominent but very divergent reworkings of this idea, through which it has disseminated widely in the humanities. Gadamer, for instance, wrote that "being that can be understood is language." What he meant is that understanding and interpretation exist primarily in language. Language, as a result, is the stuff of tradition and the medium in which people think about and relate to the world—humans can grasp and relate to nothing independently of the historical language into which they have grown. Indeed, language is the medium in which the world, as an intelligible realm, exists. Derrida, similarly, conceived of language as the carrier of intelligibility, i.e., being. The web of intelligibility in which anything that is is intelligible—in his words, "the scene of writing" (*la scène de l'écriture*) in which any text exists—is fundamentally linguistic in character. Derrida held, however, that linguistic articulations (sayings and writings) can never exhaust something's intelligibility. Something's being might be carried and articulated linguistically, but any attempt exhaustively to say/write what it is is doomed to failure: the words with which its being is articulated themselves require elucidation in additional words. Being, in short, is perpetually deferred.

The idea that all intelligibility, thinking, and action are linguistically enframed dovetails well with the kind of linguistic relativism and even idealism that had been made popular in anthropology and linguistics in the middle of the twentieth century by such thinkers as Benjamin Whorf. Although Gadamer and Derrida denied that the linguisticality of intelligibility and the world entailed a vicious relativism, the widespread dissemination of belief in the omnipresence of language that was followed from their work owes something, I believe, to the tantalizing specter of relativism. Even if I am wrong about this, the idea that language is the medium of thought/action and the world enjoyed a phenomenal career in the final third of the previous century. Its spread throughout the humanities was very much due to Gadamer's and Derrida's reception of Heidegger.

A second later Heideggerian theme that has enjoyed widespread resonance in the humanities is postmetaphysics. Heidegger, recall, described the contemporary age as the culmination of metaphysics. More specifically, the forgetfulness of being that accompanied the rise of modern technology signaled both that no further metaphysical development is possible in the West and that the Western world is condemned to endless systemization unless its people received a new understanding of being that pointed them towards new forms of life. The somewhat passive and quietist intellectual receptiveness that Heidegger thereby advocated received little subsequent affirmation (though the general mode of life he called releasement [*Gelassenheit*], or letting things be, has enjoyed considerable intellectual popularity—see below). However, the thought that the era of metaphysics is coming to a close and that the intellectual task for contemporary life is to think (and act) postmetaphysically has resonated with many subsequent theorists.

Conceptions of what qualifies as postmetaphysical thought and action vary. (For an account of what Heidegger might have meant, see Reiner Schürmann's *Heidegger on Being and Acting*.) For some, postmetaphysical thought amounts to conceptualizing human life and the world in which it proceeds in nonmetaphysical language, that is, in language uncontaminated by the attempts of metaphysicians to name presence (*Anwesenheit*—see chapter three). In the 1970s and 80s, humanist theorists debated whether a postmetaphysical language would involve the novel use of old terms or the development of novel terms, whether extant Western languages are irredeemably contaminated metaphysically, and whether any language, new or old, can stand free of the metaphysical heritage. For other thinkers, nonmetaphysical existence involves forsaking epistemological crutches. In their eyes, humans must live sans absolute truths and without assurance that the West's traditional master concepts—such as being, truth, and the good—designate stable and enduring things or matters. Positively, this tenet prescribes an ethos of attunement and responsiveness both to the contingencies, transformations, and instabilities of the particular worlds in which people act and to the inability of humans fully and reliably to grasp these worlds. A third form postmetaphysical thought is said to take is a loss of belief in essences, in the existence of a common factor that makes all entities of a given kind (e.g., bird) entities of that kind (e.g., birds). Entities of a given kind are, instead, heterogeneous. This thesis assumes special poignancy in relation to human beings, for in relation to them it becomes the doctrine that there is no common factor, no human essence or human nature, that makes humans human. What it is to be human varies among people according to their interpretations of life and their ways of living. This idea had been familiar at least since Marx, and in a way since Herder and Humboldt before him, but in humanistic quarters it acquired Heideggerian coloring in the final third of the 20th-century.

In this way, postmetaphysics segues into posthumanism. Modern thought, originating in the Renaissance, is decidedly humanistic. By "humanism" I mean a diffuse cultural sensibility that articulates the pathos of human existence and celebrates human beings as actors, creators, and thinkers. The opposition of Heidegger and others to certain types of humanism convinced many subsequent thinkers to abandon them. Three types of humanism that Heidegger, in particular, assailed are a psychological humanism that considers humans to be masters of their psyches, intentionality, and actions, an epistemological humanism that privileges the human subject-mind as the exclusive place of knowledge, and a value version that proclaims that human beings alone are responsible for their values. Today, the first of these humanisms has been widely abandoned (though hardly under Heidegger's agitation alone; names such as Marx, Nietzsche, and Freud must also be mentioned). The second humanism, meanwhile, is under strong attack (with help from other thinkers above all in science studies and philosophy of mind). The third humanism, by contrast, remains vital. Heidegger, incidentally, affirmed a fourth form of humanism that might be called "definitional humanism:" the thesis that human being is such that human life essentially contrasts with or differs absolutely from animality or mere animal life. In the late 1920s, for instance, he contrasted humans, as in-the-world, with animals construed as "world-poor" (*weltarm*) Today, definitional humanism is steadily losing adherents in the face of the contemporary life sciences.

Postmetaphysics also morphs into postmodernism. Postmodernism is an ill-advised term that nonetheless took part of the academic world by storm in the 1980s. Perhaps the best characterization of it is as a label for a confluence of ideas and stances, varying combinations of which are advocated by theorists who were standardly labeled (by others, not themselves) "postmodernists." These ideas and stances include (1) an acceptance and discernment of fragmentation, singularity, and discontinuity, (2) an unmasking of totality, reason, and theory, (3) an assault on the autonomous art-work and the sovereignty of the subject, (4) a distrust of metanarratives, and (5) a defense of pluralism, openness, and the other. Heidegger helped fuel many of these stances, though he certainly did not advocate all of them, for example, a mistrust of metanarratives. Yet, even many thinkers who affirm skepticism of metanarratives (e.g., Jean-François Lyotard) are heavily indebted to Heidegger. Indeed, later Heidegger is one of the chief intellectual sources of a plethora of "posts" (postmetaphysics, postmodernism, poststructuralism, posthumanism, even postcolonialism, though not postMarxism or postfeminism) that have marked much theory in the humanities and humanistic social sciences over the past three decades. Thereby, Heidegger has influenced thinkers too numerous to mention here.

Another area in which Heidegger' later thought has proved influential is theology. Whereas *Being and Time* made a major impression on Protestant theologians (and Heidegger's analysis of factical existence impacted New Testament scholarship), his later thought primarily made a difference to Catholic theology. In the 1930's, for example, his ideas about questioning, hearing, and being claimed were elaborated by a string of Catholic theologians, including Karl Rahner (*Spirit in the World* [*Geist in Welt*], 1957). In the 1950s, moreover, his rejection of humanism and his understanding of thinking and of language as a call were taken up by, inter alia, Heinrich Ott (*Denken und Sein*, 1959). For discussion of Heidegger and theology, see John Caputo's book, *The Mystical Element in Heidegger's Thought* (1978).

A final realm to which Heidegger has significantly contributed is environmental and ecological thought. Heidegger's vision of releasement (*Gelassenheit*), of a way of living that lets things, including things of nature, be what they are, made him a hero of the incipient German Greens in the 1950s and 60s. This vision subsequently linked up with and provided philosophical underpinnings for a variety of environmental ethics and ontologies, including versions of radical ecology, ecocentrism, and deep ecology (see Joseph Grange, "On the Way Toward Foundational Ecology" [1977], Nick Evernden, *The Natural Alien* [1985], Charles Taylor, "Heidegger, Language, and Ecology" [1985], Michael Zimmerman, "Toward a Heideggerian *Ethos* for Radical Environmentalism" [1983], and Aidan Davison, *Technology and the Contested Meanings of Sustainability* [2001]). Heidegger's impact on environmental thought also grew out of his defense of place—the place of human dwelling where people are at home—against the systematization that is rampant in the modern world of technology and tied to its objectivist, rationalist thought (see below).

The Legacy of the Spatiality of Existence

Heidegger's analysis of the spatiality of existence in *Being and Time* is an analysis of lived space understood as the space of human experiential acting. His analysis of this spatiality as orientedly (*ausgerichtet*) and dis-tancingly (*ent-fernend*) proceeding in regions of near and far placed equipment has helped foster subsequent phenomenological analyses of human spatiality in, above all, two disciplines, philosophy and geography. These are the two disciplines that appropriated his early ideas on spatiality because, *inter alia*, Heidegger decisively reshaped philosophical phenomenology and hermeneutics and geography is the social discipline most immediately concerned with space. I note that the small difference that Heidegger has made in a third discipline strongly concerned with space, namely, architectural theory, arose from his later ideas on dwelling space and not from his analysis of existential spatiality. My guess is that one reason for this selective appropriation is that architecture is concerned more with built structures than with human existence amid such structures, and in his later philosophy Heidegger shifted the center of space from human existence to things.

Although all phenomenological analyses of space after *Being and Time* are indebted to this book, it is important to distinguish two sorts of debt. One is the general conviction that spatiality is a feature of human existence (being-in-the-world). The other is the appropriation of Heidegger's specific analysis of spatiality: orientedly and dis-tancingly proceeding in regions of near and far placed equipment. The first debt can be incurred without incurring the second; indeed, this has been the pattern. After *Being and Time*, no phenomenologist doubted that his or her topics of study are features of, or contextualized within, being-in-the-world. It turned out, however, that human existence has structures and dimensions not discussed by Heidegger that bear on, or are aspects of, human spatiality.

Moreover, although Heidegger offered the first analysis of lived space, *Being and Time* was quickly followed by several prominent, though not strictly speaking philosophical, analyses of lived space, or more precisely, spatial experience that had just as much impact on subsequent phenomenological analyses of space as did Heidegger's account. Among these, Eugène Minkowski's *Lived Space: Phenomenological and Psychopathological Studies* (1933) deserves special mention. For Minkowski, lived space is, so to speak, a medium in which people necessarily live and act, a realm in which personal and collective life unfolds. The analysis of this medium contained in the last chapter of his book exerted an immense influence on subsequent phenomenological accounts of space, including those of Merleau-Ponty and Otto Bollnow. Other accounts of lived space on which subsequent theorists relied include E. Strauss "Die Formen des räumlichen Erlebens" (1930), Count Karlfried von Dürckheim, "Untersuchungen zum Gelebten Raum" (1932), and Ludwig Binswanger, "Der Raum-Problem in der Psychopathologie" (1933).

As noted, although Merleau-Ponty's famous book, *Phenomenology of Perception*, exhibits strong Husserlian traits, it is Heideggerian in its conviction that the setting of human phenomena is being-in-the-world. "The Kantian subject posits a world, but, in order to be able to assert a truth, the actual subject must in the first place have a world or be in the world, that is, sustain around it a system of meanings … [that] do not require to be made explicit in order to be exploited."[1] Heidegger's idea that existence opens a there in which entities are is retained in Merleau-Ponty's notion of the field of perception (*le champ de perception*) in which all aspects of human life are set. Human spatiality, too, is an aspect or dimension of this field, an aspect or dimension of having- or being-in-the-world: "space is existential; existence is spatial."[2] As discussed in the previous chapter, however, Merleau-Ponty emphasized the significance of the body for the human way of being-in-the-world: being-in-the-world is a bodily being-in-the-world. This thesis implies that existential structures are connected to the human body. For instance, Merleau-Ponty followed Husserl and Heidegger in viewing consciousness as riding on a sea of pre-

---

[1]   Maurice Merleau-Ponty, *Phenomenology of Perception*, trans. Colin Smith, London, Routledge, 1962, p. 129.

[2]   *Ibid.*, p. 293.

thetic, pre-reflective, and pre-objective states and ways of being. Unlike his predecessors, however, he argued that the character of these states and ways as prior is tied to their bodily nature. In his hands, moreover, human spatiality, too, becomes something bodily, to wit, the spatiality of the acting body, which is a spatiality of situatedness in the world and not a matter of position in a relational or absolute objective space. As in Heidegger, furthermore, this spatiality has two coordinated sides: a matrix of habitual action and an environment organized as a set of *manipulanda*. This division straightforwardly corresponds to Heidegger's double focus on the spatiality of being-in and the spatiality of the world. Merleau-Ponty thus provided an important correction and extension of Heidegger's analysis.

Whereas Merleau-Ponty expanded Heidegger's perspective by incorporating the body into it, Otto Bollnow espied that human existence embraces more types of space than the teleological sort that Heidegger had described. In *Mensch und Raum* (1963), Bollnow followed Heidegger in conceiving human life as necessarily spatial ("*Dasein ist räumlich*"). Humans are necessarily "space forming and space spinning out" (*Raum bildende und Raum aufspinnende*) creatures, and Bollnow remarked that being-in-the-world is almost the same as being-in-space. Bollnow distinguished, moreover, between the spatiality of human life and the lived space (*erlebt Raum*) in which humans live. The latter was conceived à la Minkowski as a medium of human life that is nothing apart from humans living in relation to it. Bollnow also joined Merleau-Ponty in emphasizing the role of the body in structuring this medium, while retaining Heidegger's account of its organization into places and regions. His lengthy analysis of lived space examines, among other topics, horizons, places, regions, distances, paths, streets, houses, doors, beds, and waking and falling asleep. Lived space, he claimed, also exhibits an inner differentiation or articulation into subspaces: hodological space (systems of paths), activity space, day- and night-space, tuned (*gestimmte*) space, presentist (*präsentische*) space, and the space of human living-together (*Zusammenleben*). Heidegger's analysis of the spatiality of existence, Bollnow held, is an account of activity space alone. It should be evident from the previous chapter that he is not entirely right in his judgment. The spatiality of existence includes some version of what Bollnow called hodological space and the space of coexistence, and it also leaves room for what he dubbed tuned space (though I did not examine this).

More recent phenomenological accounts of spatiality in philosophy follow convergent lines. In *Place and Experience* (1999), for example, the Australian philosopher Jeff Malpas, appropriating the thesis that human existence is being-in-the-world, attempts to uphold a close connection between existence, on the one hand, and spatiality and locality on the other. Being-in-the-world, he claims, is grounded in the phenomenon of place. This means that existence as it is analyzed in *Being and Time* rests on a phenomenon that received explicit attention only in Heidegger's later philosophy. More recently, Malpas has published a detailed and insightful interpretation of Heidegger's opus (*Heidegger's Topology*, 2006) that continues this line of thought. In this book, Malpas argues in detail that place, or topology, is foundational for Heidegger's thought throughout his career, and not just in phase three. This new book is the most comprehensive examination of space in Heidegger currently available. As a further phenomenological account of spatiality, I might mention my own essay "Spatial Ontology and Explanation" (1993) and book, *Social Practices* (1996). In these works, I seek to show how regionalized places devolve from teleologically organized human practices. Other phenomenological accounts of space in philosophy include H. Lassen's *Beiträge zu einer Phänomenologie und Psychologie der Anschauung* (1939), Elisabeth Ströker's *Philosophische Untersuchungen zum Raum* (1965), Walter Gölz's *Dasein und Raum* (1970), Lenelis Kruse's *Räumliche Umwelt* (1974), and Bernhard Waldenfels's *In den Netzen der Lebenswelt* (1985).

The other discipline which took up Heidegger's analysis of the spatiality of human existence is geography. Perhaps not surprisingly, those propounding or building on his analysis have primarily been of a phenomenological persuasion. In an oft-cited article, "Grasping the Dynamism of Lifeworld" (1976), Anne Buttimer aims to bring together two approaches to space and geographical thought that are often thought to be antagonistic: phenomenology and typological/functional analysis. Central to this rapprochement are the phenomenological concepts of lifeworld, body subject, and intersubjectivity. Buttimer also shows how upholding Heidegger's intuition that person and world are inseparable illuminates such geographical matters as sense of space, social space, ecological processes, and functional organization. John Pickles's 1985 book, *Phenomenology, Science and Geography*, offers the most complete presentation in geography of Heidegger's views on phenomenology, science, the human sciences, human existence, and space. Pickles discusses how the unexamined objectivist ontology that underlies the history of geography has colored the reception of phenomenology in that discipline. He draws extensively on *Being and Time* in challenging geographic objectivism (as well as the objectivist character of the human sciences more broadly) and in locating the source and genesis of both objectivism and objectivist theory in the existence of the individual theorist. The book concludes with Heideggerian accounts of human science as an interpretive enterprise and of the spatiality of human existence, the subject matter of geography. Pickles argues that place and space as geographers discuss them are related, that both derive from a more fundamental spatiality, namely, the spatiality of human existence, and that grasping both this spatiality and how objective place and space are grounded in it requires an existential interpretation of nature, environment, and world.

A third geographical reception of Heidegger's existential spatiality is found in Edward Soja's *Postmodern Geographies* (1989). Soja's spatial ontology is strongly rooted in the spatial ideas of the English sociologist Anthony Giddens. In *Central Problems in Social Theory* (1979), Giddens argues that human activity occurs in locales. To speak of a locale is to speak of the use of space as the setting of interaction. It follows that the identities of such locales as rooms, street corners, towns, and national territories derive from the routinized social practices that transpire within and use them. That a locale is a kitchen, for instance, derives from the routinized practices that transpire in it, as does the organization of entities there. This is a very Heideggerian intuition because social practices are conceived of as instituting the locales, that is, the spaces, where they themselves transpire; like Dasein, as Bollnow wrote, social practices are spatial (in Giddens's words, practices "regionalize" space). Soja preserves this idea. Much more than Giddens does, however, he takes to heart Heidegger's intuitions that spatiality is a dimension of being-in-the-world and that being-in-the-world is inherently spatial. He describes spatiality as a "multilayered system of socially created nodal regions, a configuration of differentiated and hierarchically organized locales. The specific forms and functions of this existential spatial structure vary...but once being is situated in-the-world the world it is in becomes social within a spatial matrix of nested locales."[3] (A node is a clustering of activities around identifiable geographical centers.) This analysis underlies a general account of the spatiality of social phenomena. Pickles' and Soja's ontologies can be conjoined to form a comprehensive account of human spatiality that ranges from the spatiality of ongoing activity to the spatial structures that mark advanced capitalist societies.

To conclude this discussion of the legacy of existential spatiality, I mention two social scientific appropriations of it outside of geography, both in archaeology. In *Time, Culture and Identity* and *Social Being and Time*, respectively, the English archaeologists Julian Thomas and Christopher Gosden develop Heideggerian interpretations of human life. Thomas's account is particularly indebted to Heidegger. Extending Heidegger's ontology of regions and places, Thomas analyzes the production of places by reference to the human inhabitation of spatial locales. He is one of the few Heideggerians to take up the phenomena of orientation and dis-tance, though he follows Merleau-Ponty in tying them to the body. Thomas also appropriates Heidegger's later terms "building" and "dwelling," conceptualizing them in line with the account of existential spatiality found in *Being and Time*. Whereas dwelling is being at home in one's surroundings in the sense of going about one's business in familiarity with them, building is transforming the world so that a location becomes the embodiment of a particular way of existing in the world. Gosden, meanwhile, offers a more neo-Heideggerian picture of space. He defines spaces as material arrangements such as landscapes and dwellings and argues that spaces, and also times, are a dimension of networks of action. Material configurations are features of action networks because they are arrangements amid which humans act: they are products of such networks that in turn determine these networks. Spaces determine actions in (1) constituting open areas through which people can move, (2) setting bounds on movement, and (3) acting as stage settings for human interaction. Gosden's account articulates a Heideggerian sense of theembeddedness of activity in the world, though it is considerably more materialist than is Heidegger's account and fails to theorize how

---

[3]   Edward Soja, *Postmodern Geographies: The Reassertion of Space in Critical Social Theory*, London, Verso, 1989, p. 148.

the world gets taken up into activity (cf. Heidegger's notion of equip-ment).

The Legacy of Place and Dwelling

As explained in the previous chapter, Heidegger's later thoughts on space concern a type of lived space different from that analyzed in *Being and Time*. Whereas the latter tackles lived space understood as the space of experiential acting, the space of humans experientially proceeding amid entities, later works examine lived space understood as spaces of living—spaces with which living is involved, in terms of which it proceeds. Whereas the space of experiential acting is under-stood as that *wherein* human life proceeds, spaces of living are con-strued as *where* it does so. To analyze this where, moreover, new terms of art are marshaled, including place (*Ort*) and site (*Stätte*). A place is a place of being; it is where a clearing opens. A clearing, moreover, opens at one or more particular entities in the clearing, which Heidegger called "sites." Examples are works of art, great po-litical deeds, the polis, features of the natural and built environment, and language. Most important for the present chapter, sites of being are also sites of dwelling. Sites admit, or make room for (*einräumen*), a space of human dwelling and accomplish this by way of setting out places and paths at and along which humans live daily. As entities at which clearings open, sites, or things, simultaneously admit (*ein-räumen*) the fourfold. Indeed, things admit the fourfold by setting out paths and places where humans dwell. In other words, a space of dwelling is automatically the open of a clearing: a space of dwelling is at once an open in which entities are (more precisely: a dwelling time-space is at once an open in which entities are). Humans, that is, neces-sarily stand into the clearing.

These details of Heidegger's later thoughts on space have found lit-tle direct resonance in later spatial theory. That later spatial theorists ignored his account of the clearing is not surprising given its obtuse-ness. But they have equally ignored his more concrete idea that things set out spaces for human habitation. About the only serious appro-priation of this idea has been in architectural theory. In *Elements of Architecture: From Form to Place* (*De la Forme au Lieu*, 1986), for example, the Swiss theorist Pierre van Meiss claims that places give spaces and that most every building is an occasion to make a place and add to the continuous articulation of the human habitat. Similarly, the renowned Norwegian theorist Christopher Norbert-Schulz wrote of how buildings afford spaces for human life. In "Point de Folie— Maintenant L'Architecture" (1986), finally, Jacques Derrida conceptualized buildings as events, as takings place: buildings are givings that give places to themselves and give spaces for things to happen. Generally speaking, however, this provocative Heideggerian line of thought has been overlooked.

To make this, however, the final word on the contribution of later Heidegger's philosophy to spatial thought would be seriously to misrepresent it. For a broader intuition that animates Heidegger's later thoughts on space has and continues to inform much spatial theory. This intuition is that humans dwell in places: that places are where human life transpires and that human life essentially involves dwelling in them. This intuition kindles spatial thought in contemporary architectural theory, geography, philosophy, archaeology, and anthropology, even though theorists vary as to whether they closely follow or widely stray from Heidegger's particular articulation of it. Of course, the notion of place, like the distinction and sometimes opposition between place and space, is pervasive in contemporary spatial thought. Heideggerian theories form just one family of accounts of it. Often, moreover, Heideggerian conceptions of dwelling and place join with other conceptions of them to jointly inform theories that make dwelling central to the analysis of spatial practices or social space as products and constituents of society (e.g., the work of Michel Lussault, Jacques Lévy, Augustin Berque, and Kirsten Simonsen; see bibliography). Other Heideggerian themes that appear widely in contemporary spatial thought are the coordination of building and dwelling and the distinction between authentic and inauthentic places, buildings, or ways of life.

The discussion below of the legacy of Heidegger's conceptions of place and dwelling is confined to those thinkers who explicitly appropriate and work with them. I should mention, therefore, that a wealth of prominent thinkers such as Emmanuel Lévinas, Henri Lefebvre, Jacques Derrida, Luce Irigaray, and Jean-Luc Nancy have interrogated Heidegger's conceptions of place and dwelling or made these conceptions the starting points of their own reflections on various topics, for example and respectively, openness to alterity, forms of space, hospitality, sexual difference and gendered life, and community (see bibliography for references). Of greatest relevance for the present book is the close correspondence between dwelling and place, on the one hand, and what Lefebvre called representational space, or more perspicuously, lived space, on the other (see *The Production of Space*, 1974). According to Lefebvre, representational space, together with spatial practice and representations of space (respectively, perceived and conceived space) are the three forms of space pertinent to human life. Lefebvre also agreed with Heidegger that objectivist conceptions of space are tied to the systematic ordering and mastering characteristic of the age of technology. In being taken up or reacted to by leading representatives of contemporary French thought, Heidegger's understandings of the significance of topology, of dwelling and place, for human life has diffused well beyond those theorists who explicitly appropriate and apply his conceptions of these phenomena.

Norberg-Schulz propounded perhaps the purest Heideggerian view on space today. This view is laid out in a series of books, *Existence, Space, and Architecture* (1971), *Genius Loci: Toward a Phenomenology of Architecture* (1979-80), and *The Concept of Dwelling* (1985). Humans dwell, Norbert-Schulz maintained, when they can orient themselves within and identify with an environment. Orientation is knowing where you are, whereas identification is knowing how you are a certain place. Whereas orientation embraces such phenomena as centers, paths, distances, and directions, identification centers on the feeling of being at home. Humans dwell when they belong to places, which are distinctive spaces that offer "existential footholds" for people. The distinctiveness of a place, moreover, is called its genius loci—the specific constellation of meaningful things that gather or are gathered to form it. The meanings of the things involved are relative to human activity and constitute the matrix within which people proceed in the places concerned. Dwelling, in short, is the appropriation of a world of meaningful things.

The function of architecture is to create places through the construction of built structures at which they are anchored. Its products are a type of poetry that make geniuses loci concrete, gather things and lives, and thereby "help man dwell poetically." To analyze this situation, moreover, Norberg-Schulz freely used Heidegger's concepts, including the Hölderlinian ones of earth, sky, mortals, and gods.

The world which is gathered by a work of architecture is...an "inhabited landscape," that is, a landscape which has been "understood" as a particular case of the totality earth-sky, in relation to the four modes of dwelling. The works of architecture...make this understanding a concrete fact. As things, they fulfill their gathering function through their bodily forms.[4]

Norberg-Schulz further claimed that contemporary Western life suffers from a loss of place and from the spread of monotony and a sense of nowhere. Too often, he wrote, attention today focuses solely on orientation at the expense of identification, which is left to chance. The result is that built environments no longer afford places and rests for dwelling, where dwelling is construed as being at peace in a protected place. This sort of analysis of contemporary society is widespread among Heideggerian spatial theories. Inspired, above all, by Heidegger's account of the age of technology, according to which human life is increasingly being incorporated into social systems organized around mastery and ordering, these theories typically describe living in place as a redemptive rooted life that opposes the unattached anonymous existence that prevails in the expanding rationalist and objectivist spaces of the capitalist system. It would not be terribly inaccurate to label this sensibility "neo-romantic." Norberg-Schulz also called on the above ideas to ground theories of the different types of dwelling (settlement, urban space, institution, and house) and of the character, "language" (morphology, topology, and typology), and history of architecture. Other Heideggerian understandings of dwelling and place in architectural theory can be found in van Meiss's *De la Forme au Lieu* and Karsten Harries's *The Ethical Function of Architecture* (1997). Harries, for instance, draws on both Heidegger's essay on the origin of the artwork and his ideas about the relation between building and dwelling to argue that the ethical function of architecture is to articulate a people's ethos: to give form to a people's identity and to speak to them about how to live.

Geography is a second discipline that has made copious use of a Heideggerian pairing of dwelling and place. One prominent representative is the Canadian geographer Edward Relph. At first an advocate of the Husserlian idea that the task of phenomenology is to explore consciousness and subjectivity, Relph subsequently warmed to later

---

[4]   Christian Norberg-Schulz, *The Concept of Dwelling: On the way to figurative architecture*, New York, Electra/Rizzoli, 1985, p. 19.

Heidegger. His *Place and Placelessness* (1976) focuses on place as an aspect of the lived world, as a feature of the existential and perceptual spaces of lived experience. Among the aspects of places that he examines are location, time, community, rootedness in place, and home and identity. Drawing on ideas from *Being and Time*, he also contrasts authentic and inauthentic senses of place and place-making. An authentic sense of place is "being inside and belonging to *your* place both as an individual and as a member of a community, and to know this without reflecting on it."[5] An inauthentic sense of place, by contrast, is lacking an appreciation of the significance of place and belonging to places as anyone does (cf. the discussion of *das Man* in chapter three). Authentic place-making, meanwhile, requires a "clear and complete conception of man as well as a sensitivity to the significance of place in everyday life,"[6] whereas its inauthentic counterpart involves building (1) on the basis of both technique and conceptions of objective space and (2) in the absence of an understanding of the being of human existence and its relation to place. Similar to Norberg-Schulz, Relph claims that the contemporary world is marked by placelessness. This is a condition in which people have lost appreciation for the significance of places, and the identities of places are so weakened that "they not only look alike but feel alike and offer the same bland possibilities for experience."[7]

This Heideggerian approach continues in *Rational Landscapes and Humanistic Geography* (1981). In this book, Relph defends what he calls "environmental humility." This way of life combines guardianship of the individuality of places, insight into what environments and situations are, and a building practice that is responsive to local environments and the needs of local users. The way of life that contrasts with environmental humility is controlling environments and communities and subjecting them to the imperatives of rationalist planning and building. Environmental humility is clearly a version of Heidegger's releasement (*Gelassenheit*): letting places and communities be and tending to as well as guarding them in their identity. In Relph's hands, it is also the basis of an environmental ethics.

---

[5]   Edward Relph, *Place and Placelessness*, London, Pion, 1976, p. 65, italics in original.

[6]   *Ibid.*, p. 71.

[7]   *Ibid*, p. 90.

A second prominent geographical proponent of a Heideggerian joining of dwelling and place is the American geographer and architectural theorist David Seamon. In his book, *A Geography of the Lifeworld* (1979), Seamon describes geography as the study of the earth understood as man's dwelling place. The book focuses on everyday geographical experience, the sum-total of a person's involvement with the spatial world in which she lives. Seamon appropriates Merleau-Ponty's conception of a body subject to analyze human activity as body ballets and bodily time-space routines. Drawing on features of lived space such as centers and places for things, he also holds that place-ballets are key to the phenomena of dwelling at and being at-home in places. In addition, Seamon, like Relph, deploys the contrast between authenticity and inauthenticity to describe the fate of places in the contemporary world.

Seamon has continued to promote Heideggarian and phenomenological perspectives on space in several anthologies, most notably *Dwelling, Place & Environment* (1985, with Robert Mugerauer) and *Dwelling, Seeing, and Designing* (1992). The first of these volumes brings together architects, geographers, philosophers, psychologists, and other theorists of a generally phenomenological persuasion. As the title indicates, the book is strongly Heideggerian in approach and topic. Among the essays that directly discuss or invoke Heideggerian themes and ideas, Edward Relph explores the Heideggerian underpinnings of geography, the architectural theorist Kimberly Dovey theorizes authentic environmental meaning and the task of building authentic places, Robert Mugerauer examines the significance of language in human encounters with the world in the context of an environmental hermeneutics, and philosopher Joseph Grange studies the multifaceted exchange, engagement, and mutual openness between body and environment.

The second volume focuses on the built environment. An initial trio of Heideggerian essays by Edward Relph, Karsten Harries, and the American environmental designer Catherine Howett argues that modern Western landscapes are shaped by economic, technological, and aesthetic concerns that reflect neither a full understanding of human existence nor the contribution of place and dwelling to human identity, perception, and understanding. Several of the remaining essays offer either Heideggerian investigations of specific environmental or architectural phenomena (landscapes, porches) or Heideggerian-informed reflections on the design, creation, and understanding of community. Overall, the book outlines what Seamon calls "phenomenological ecology," whose task is to describe how things, life forms, people, events, and situations come together environmentally.

In this context, I might also mention the American architectural theorist Robert Mugerauer's *Interpreting Environments* (1995), which goes beyond phenomenological accounts of the environment to examine deconstructive and hermeneutic approaches to it as well. Mugerauer's main exemplar of deconstruction is Derrida, whereas his main exponent of hermeneutics is Gadamer. This book thus charts Heidegger's further-flung significance for spatial thought.

Dwelling and place are also central to a fine essay by the English geographers Paul Cloke and Owain Jones, "Dwelling, Place and Landscape: an Orchard in Somerset" (2001). Characterizing dwelling à la Heidegger as embracing the "rich intimate ongoing togetherness of being and things which make up landscapes and places," they argue that the concept offers insights into how human activities are embedded in landscapes, places, and networks; how nature and culture intertwine in place; and how all this joins past, present, and future. Dwelling, they claim, embraces an embodied and imaginative rootedness in place, thus a correspondence among community, landscape, and place. Like Relph and Seamon, Cloke and Jones take over Heidegger's notion of authenticity, claiming that once the romanticist overtones of the concept of dwelling are overcome, and the concept is adapted to the contemporary world, authentic dwelling remains possible. As a sort of geographical counterweight to Cloke and Jones's affirmation of Heidegger, I might mention an essay by Paul Harrison titled "The Space Between Us: Opening Remarks on the Concept of Dwelling" (2007). This essay explicates Lévinas's claim that Heidegger's conception of dwelling fails to acknowledge as essential to the phenomenon the openness to the truly other that Lévinas claimed is key to human existence. Harrison suggests that rethinking dwelling so as to incorporate this openness into it requires reconceiving the dynamics of subjectivity and subject formation in interpersonal encounters.

A final Heideggerian geography text worth mentioning in the present context is *Mapping the Present* (2001), by the English geographer Stuart Elden. This book does not so much theorize space as study the relationship between Heidegger's and Foucault's historical and spatial approaches and the bearing of these approaches on the general relationship between space and history. Elden's overall point is that history should be written as spatial history, as a story of matters that are inherently spatial in character. The first half of the book is an interpretation of Heidegger's ideas on space and what else in his philosophy they bear on. The second half claims that seeing Foucault's work through a Heideggerian lens reveals that Foucault's genealogies, which he, Foucault, called "histories of the present," are more clairvoyantly labeled "mappings of the present." This means that Foucault's histories are spatial histories in the sense that space is basic to their conceptualization of their subject matters. As such, they serve as a model for the propitious writing of history. (On Heidegger and Foucault, see also Hubert Dreyfus, "On the Ordering of Things: Being and Power in Heidegger and Foucault," 1992.)

Heidegger's ideas on dwelling and place made their way to anthropology and archaeology in the 1990's (for an overview of place in archaeology, see Julian Thomas, "Archaeologies of Place and Landscape" [2001]). I already mentioned that Thomas' book, *Time, Culture, and Identity*, appropriates Heidegger's concepts of dwelling and place but interprets them in line with the analysis of existence in *Being and Time*. In the early 1990's, the prominent English anthropologist Tim Ingold influentially reworked the concept of dwelling. For Ingold, to dwell is to inhabit. To adopt a "dwelling perspective" on phenomena such as landscapes (understood to include natural terrain, the built environment, and their spaces) is to consider the world and its organization as it is disclosed to those who inhabit it. Following Merleau-Ponty and Dreyfus's interpretation of *Being and Time*, Ingold analyzes inhabiting as skillfully carrying out practical activities with and amid entities. This analysis, in turn, underwrites the concept of a taskscape as the ensemble of tasks that are carried out in given dwelling practices. Landscapes arise from human activity; it follows that landscapes and taskscapes are correlated. Indeed, activity, landscape, and taskscape form an indissolvable whole, and to speak of dwelling is to approach human activity as an inherent component of this whole. One implication of this idea is that people, in dwelling, do not act on the world. Rather, they move along with, that is, are part of the world's continuous transformation: activity streams are part of the unfolding whole composed of activity, landscape, and taskscape.

Accordingly, to consider landscapes or buildings from the dwelling perspective is to consider the environment—its organization and construction—as part of this whole, thus as inherently tied up with human practices. Approaching landscapes this way means, among other things, treating landscapes as embodiments of taskscapes. Doing this also means abjuring the usual practice of thinking that built components of the landscape are made before they are inhabited and instead treating built structures as arising in the current of human practices, as crystallizations and extensions of ongoing human activity that transpire in already meaningful settings. The embeddedness of buildings in ongoing activity suggests that they must reflect and be suitable for living before they are constructed. This claim is an interpretation of Heidegger's oft-cited remark that "Only if we are capable of dwelling, only then can we build."[a] (BDT 160)

Ingold's analysis synthesizes ideas from Heidegger's second and third phases. It has also been appropriated by various social theorists, including Cloke and Jones, the English archaeologist Christopher Tilley (*A Phenomenology of Landscape* [1994]), and the English sociologists Phil Macnaghten and John Urry in their influential account of the intersection of nature and society in *Contested Natures* (1998).

Tilley's book represents a synthesis of much of the above. Characterizing phenomenology as a mix of Heidegger and Merleau-Ponty, he draws on Pickles, Relph, Seamon, and Seamon's coedited volumes to develop a phenomenological account of space, place, and landscape. He also applies this account to three prehistoric landscapes in Wales and Southwest England. Ingold, moreover, is brought in both to conceptualize people and environment as constituents of a single whole and to treat landscape as the medium for, and outcome of, human practices.

Philosophers, finally, have also drawn on Heidegger's notions of dwelling and place. Following Heidegger's fleeting characterization in *Being and Time* of being-in as dwelling amid (*wohnen bei*, SZ 54), several philosophers, for example Merleau-Ponty and Bollnow, have defined the relationship of humans to the world in which they exist as dwelling. Others have taken up Heidegger's later notion of dwelling and analyzed the forms of human dwelling. Notable among them is the French philosopher Gaston Bachelard, though it would be misleading to describe Bachelard as Heideggerian (his texts are too original and draw on diverse sources, many nonphilosophical). According to Bachelard, dwelling concerns the inhabitation of the intimate spaces of human life. His 1958 book, *The Poetics of Space* (*La poétique de l'espace*), focuses, accordingly, on the house and its spaces. It analyzes the house as such as well as its forms, the "houses" of hidden things (e.g., drawers and closets), the places of daydreams, small places where people curl up, the dialectics of small and large and of inside and outside, and roundness. In all instances, these matters are treated as images and not as objective facts, thus as closely bound up with people's inner life. Indeed, Bachelard's "topoanalysis" is an account of the places of dwelling and, at one with this, the inner life of dwellers. To carry it out he wove together phenomenology, psychoanalysis, psychology, and literature. Another philosopher of space inspired by Heidegger, among other thinkers, is Edward Casey. In *Getting Back into Place* (1993), Casey develops an account of being-in-place, which he claims is a more concrete conception of human existence than is being-in-the-world. Prominent among the book's many sources are Merleau-Ponty, whom Casey draws on in theorizing the role of the body in implacement and displacement, and Heidegger, whom Casey thanks for making place an issue and appropriates in conceptualizing both how humans dwell in places and what it means to build them.

Philosophers have also joined those critical of Heidegger's account of dwelling and place for its neo-romanticist invocation of unified peoples living in particular, archetypically rural places. One prominent such critic is Emmanuel Lévinas, who noted that Heidegger's emphasis on social unity (see the previous chapter) perpetuates the pervasive idea in Western thought that communion is the highest form of social relation. Another, contemporary critic is the American philosopher David Kolb. In his 1990 book, *Postmodern Sophistications*, Kolb agrees with Heidegger (and many others) that the contemporary world needs places as opposed to more spaces, but challenges Heidegger's—the West's—presumption that the ideal social arrangement is a unified people living in particular discrete places (for versions of this criticism in geography, see David Harvey, *Justice, Nature, and the Geography of Difference* (1996) and Nigel Thrift, "Steps Toward an Ecology of Place" [1999]). Such a vision underplays the coexistence in the contemporary world of diverse people at close quarters and how places for millennia have interpenetrated and shaped one another. What is needed today are places that, while drawing on local cultural as well as architectural languages and traditions, affirm interdependence, accommodate multiplicity, and open up possibilities for everyone (and different possibilities for different people). On this issue, see also the English anthropologist Barbara Bender's *Stonehenge: Making Space* (1998). Bender affirms a Heideggerian approach to place and space, but criticizes Heidegger for emphasizing rootedness in rural activities, past times, and "simple oneness." She also argues that combining Heidegger's approach with Raymond's Williams' concept of structures of feelings (*The Long Revolution*, 1961) allows the theorist of space to take social, economic, and political contexts into account.

A response of sorts to Kolb, Bender, and others can be found in Robert Mugerauer's *Interpretations on Behalf of Place* (1994). Mugerauer argues that Heidegger's emphasis on rootedness need not be interpreted neo-romantically. Moreover, he claims, Heidegger's approach to place and dwelling contains resources sufficient both to accommodate the postmodern emphasis on multiple lifestyles and meanings and to secure a type of environmental restoration that combines innovation with the retrieval of traditions. Different types of response to the accusation of romanticism can be found in Albert Borgmann's *Technology and the Character of Contemporary Life* (1984) and Charles Spinosa, Fernando Flores, and Hubert Dreyfus' *Disclosing New Worlds* (1997).

The Legacy of the Clearing

Social ontologies are accounts of the nature or basic features of social life. The social theoretical legacy of Heidegger's notion of the clearing lies, not in accounts of human spatiality, but in particular social ontologies. The present section provides an overview of the most consequential Heideggerian social ontologies. Joining them is the thesis that a space, in most cases, an open of being and intelligibility, is central to the constitution of the social or of some key dimension of it.

I should first describe one immensely influential philosophical appropriation of the clearing that is not conceived of as (part of) a social ontology. In section one, I mentioned Derrida's concept of the web of intelligibility, in his words, the concept of the scene of writing. The scene of writing, the wherein or where of all textuality (all articulation of intelligibility, all entities), is an interpretation of Heidegger's clearing. Derrida's infamous phrase, "there is nothing outside the text," simply means (1) that to be is to be intelligible and, in this sense, to be a text and (2) that all articulations of intelligibility, that is, all entities occur in the scene of writing. This scene is prior to any time and space that belongs to entities. Indeed, it is where all such times and spaces occur. Its basic action, moreover, is spacing (*espacement*), the opening of possible determinacies through the inauguration of differences. Notice that the scene of writing, like the clearing, is a prespatial space. Derrida's notion of a linguistic space of being has guided many post-structuralists after him and represents an important dimension of Heidegger's legacy.

Of more direct importance to social thought are two other philosophical interpretations of the clearing. According to the Canadian philosopher and political theorist Charles Taylor, social reality is practices (see "Interpretation and the Sciences of Man" [1971]). The actions and acts of speaking and writing that compose a practice articulate a "semantic space," a space of intelligibility, that envelops everything that henceforth composes or is encountered in the practice. Practices, in other words, open fields of meaning in terms of which they themselves proceed. The individuals who carry on practices, as well as the relations they maintain in doing so, are embedded in the fields of meaning that the practices they carry on open and sustain.

A similar picture is found in Spinosa, Flores, and Dreyfus's *Disclosing New Worlds* (1997). The authors define a world (also called a "disclosure space") as an organized set of practices that produces a self-contained web of meanings. Any such world is organized by an array of activities, purposes, equipment uses, and personal identities, and also by an embracing style of how things matter to people. People mostly carry on their lives in particular worlds, meaning that what they do and who they are is beholden to the meanings and styles that organize these worlds. The realm of the social is a field of hierarchically organized worlds (disclosure spaces).

Two other social ontological appropriations of Heidegger's clearing akin to the two just discussed are Joseph Rouse's account of scientific practices (*How Scientific Practices Matter*, 2003) and my conception of the site of the social—the context as part of which social life inherently transpires (*The Site of the Social*, 2002). Neither Rouse's nor my analysis treat the clearing primarily as a space of intelligibility. Instead, whereas Rouse conceptualizes it as an interactional field within which distinct entities (e.g., objects and people) crystallize, I treat it as the constitutive context of which social life is inherently a part. This context, I argue, is composed of nexuses of practices and material arrangements.

Taylor and the team of Spinosa, Flores, and Dreyfus offer social ontologies that make the clearing qua space of intelligibility central to the constitution of the social. Parallel conceptions are found in the work of the French sociologist Pierre Bourdieu and the coproductions of the English political theorist Ernesto Laclau and the French philosopher and political thinker Chantal Mouffe.

Like the ontologies just canvassed, Bourdieu's social ontology highlights spaces that combine activity and meaning. Central to his account of social life is the notion of a field (see *Outline of a Theory of Practice* [1972] and its sequel *The Logic of Practice* [*Le sens pratique*, 1980]). A field is a bounded realm of activity; examples are agriculture, politics, marriage, and education. The actions and entities found in a given field boast interrelated meanings, which arise from the batteries of action dispositions—"habitus"—that generate activity in that field and are correlated with the objective statistical regularities that characterize it. Certain matters, moreover, are at stake in a given field's activities (e.g., profit, prestige), and actors draw on multiple capitals (cultural, symbolic, economic) in pursuing these stakes. All in all, a field comprises stakes, a distribution of capitals, and a space of actual and possible activities and meanings. Bourdieu also contended that different fields are organized homologously and, together, compose larger wholes called societies. For Bourdieu, consequently, the social is an expanse of fields containing extensive continents of homologous activity and meaning spaces.

Central, meanwhile, to Laclau and Mouffe's analysis of social affairs in *Hegemony and Socialist Strategy* (1985) is the concept of a discourse. A discourse, for them, is not a pure linguistic entity; it is, instead, an array of interrelated and systematically meaningful actions, things, and words. These discourses are the regimes of order that compose social life. According to the authors, the sum-total of discourses, or regimes of social order, that exist over any period of time subsides in something they call the "field of discursivity." This field is an inexhaustible openness of articulability that overflows the totality of determinacies contained in the sum of discourses at any time. This inexhaustibility undercuts the ostensible stability of any discourse in the field. The field of discursivity is clearly a version of Derrida's scene of writing, thus of Heidegger's clearing. According to Laclau

and Mouffe, social life is organized as a sum-total of systematic con-
stellations of meanings that subsists in an inexhaustible field of potential meaning.

The German-American philosopher and political theorist Hannah
Arendt made a very different sort of social theoretical appropriation of
Heidegger's clearing. In her epochal *The Human Condition* (1958),
she described the public sphere as an open space in which people appear to one another amid a world of things in common. It is a space of
mutual visibility that automatically opens when people exist in one
another's presence. This public space also contrasts with people's
intimate thoughts and feelings, which lead a "uncertain, shadowy
existence" in the "twilight" of private subjective experience. Arendt
did not claim of the public, as Heidegger did of the clearing, that it is
constitutive of human existence. Unlike, moreover, the phenomena
Heidegger dubbed being-with and *das Man*, Arendt's public requires
the copresence of human beings and does not construe people as the
same. Nonetheless, Arendt inherited from Heidegger the sense of an
unavoidable open space that embraces different individuals who appear in it. In *In Defense of Politics* (1962), the English political theorist
Bernard Crick appropriated Arendt's public as the basis of an unconventional theory of politics, government, and rule.

Finally, in his essay "The event of space" (1998), the German geographer Ulf Strohmayer both critiques and appropriates conceptions of
space found in Heidegger and other phenomenologists. Discussing
the epistemological problems that attend any treatment of the clearing
as the constitutive background of human knowledge (where the clearing is understood à la Gadamer and Derrida as a space of language and
tradition), Strohmayer argues that space should be theorized through
Heidegger's notion of the event (*Ereignis*). Theorizing space through
the event is a relatively new idea in spatial theory. Following *Contributions to Philosophy*, Strohmayer claims that approaching space
through the event involves thinking the "a priori linearity of time and
space as absent ground [*Abgrund*]…or as 'play'…rather than as
'place.'"[8] Central to this effort is theorizing the materiality of the event
of space.

---

[8]   Ulf Strohmayer, "The event of space: geographic allusions in the phenome-
      nological tradition," *Environment and Planning D: Society and Space* 16
      (1998): 105-21, here 118.

The idea of theorizing space through the event has arisen concomitantly with increased scholarly attention to Heidegger's *Contributions to Philosophy*. My sense is that theories of space developed by scholars who draw on later Heidegger will increasingly take this form. I also believe that greater attention to the *Contributions* will turn spatial theory toward the phenomenon of timespace. In the previous chapter, I explained that Heidegger, in his third phase, interpreted the clearing as a temporalspatial affair and not as a spatial (or temporal) one alone. Interpretations and appropriations of Heidegger's views on time and space have focused on time or space alone, a practice that reflects the widespread tendency in social thought to treat time and space separately. The timespace structure of social life that is implicated in Heidegger's later account of the clearing remains to be theorized.

6 Selected Significant Writings and Secondary Literature

Works Published by Heidegger

GA = *Collected Works* (*Gesamtausgabe*)

Die Lehre vom Urteil im Psychologismus. Ein kritisch-positiver Beitrag zur Logik [*The Theory of Judgment in Psychologism. A Critical-Positive Contribution to Logic*], Leipzig, Johann Ambrosius Barth, 1914, also in GA 1, Frankfurt a/M, Vittorio Klostermann, 1978, pp. 59-188.

Die Kategorien- und Bedeutungslehre des Duns Scotus [*The Theory of Categories and Meaning in Duns Scotus*], Tübingen, J.C.B. Mohr, 1916, also in GA 1, Frankfurt a/M, Vittorio Klostermann, 1978, pp. 189-411.

Being and Time, trans. John Macquarrie and Edward Robinson, New York, Harper & Row, 1978; *Sein und Zeit*, fifteenth edition, Tübingen, Max Niemeyer, 1979 [1927].

Kant and the Problem of Metaphysics, trans. James S. Churchill, Bloomington, Indiana University Press, 1962; *Kant und das Problem der Metaphysik*, GA 3, Frankfurt a/M, Vittorio Klostermann, 1973 [1929].

Elucidations of Hölderlin's Poetry, trans. Keith Hoeller, Amherst, Humanity Books, 2000; *Erläuterungen zu Hölderlins Dichtung*, GA 4, Frankfurt a/M, Vittorio Klostermann, 1981 [1936–68].

Off the Beaten Track, trans. Julian Young and Kenneth Haynes, Cambridge, Cambridge University Press, 2002; *Holzwege*, GA 5, Frankfurt a/M, Vittorio Klostermann, 1980 [1935–46].

Vorträge und Aufsätze, Pfullingen, Günther Neske, 1954 [1936–53], also GA 7, Frankfurt a/M, Vittorio Klostermann 2000; essays published in *Poetry, Language, Thought*, trans. Albert Hofstadter, New York, Harper & Row, 1971 ("Building Dwelling Thinking," "The Thing," and "...Poetically Man Dwells..."), *The Question concerning Technology and Other Essays*, trans. William Lovitt, New York, Harper & Row, 1977 ("The Question Concerning Technology" and "Science and Reflection"), *Early Greek Thinking*, trans. David F. Krell and Frank A. Capuzzi, New York, Harper & Row, 1975 ("Logos (Heraclitus, Fragment B 50)," "Moira (Parmenides VIII, 34-41)," and "Aletheia (Heraclitus, Fragment B 16)"), *Nietzsche II: The Eternal Recurrence of the Same*, trans. and ed. David F. Krell, New York, Harper & Row, 1984 ("Who is Nietzsche's Zarathustra?"), and *The End of Philosophy*, trans. Joan Stambaugh, New York, Harper & Row, 1973 ("Overcoming Metaphysics").

Pathmarks, William McNeill (ed), Cambridge, Cambridge University Press, 1998; *Wegmarken*, GA 7, Frankfurt a/M, Vittorio Klostermann, 1973 [1919–61].

Principle of Reason, trans. Richard Lilly, Bloomington, Indiana University Press, 1977; *Der Satz vom Grund*, Pfullingen, Günther Neske, 1957, also GA 10, Frankfurt a/M, Vittorio Klostermann, 1997.

"*Hebel—Friend of the House*," trans. Bruce V. Foltz and Michael Heim, in *Contemporary German Philosophy*, Volume 3, Darrel E. Christenson et al. (ed), University Park, The Pennsylvania State University Press, 1983; *Hebel, der Hausfreund*, Pfullingen, Günther Neske, 1957.

Identity and Difference, trans. Joan Stambaugh, New York, Harper & Row, 1969; *Identität und Differenz*, Pfullingen, Günther Neske, 1957, also GA 11, Frankfurt a/M, Vittorio Klostermann, 2006.

Discourse on Thinking, trans. J.M. Anderson and E.H. Freund, New York, Harper & Row, 1966; *Gelassenheit*, Pfullingen, Günther Neske, 1959 [1944–55].

On the Way to Language, trans. Peter D. Hertz, New York, Harper & Row, 1971; *Unterwegs zur Sprache*, Pfullingen, Günther Neske, 1959 [1950–59], also GA 12, Frankfurt a/M, Vittorio Klostermann, 1985.

Nietzsche I, Pfullingen, Günther Neske, 1961 [1936–39], also GA 6.1, Frankfurt a/M, Vittorio Klostermann, 1996. Only one part translated in *Nietzsche III: The Will to Power as Knowledge and Metaphysics*, trans. Joan Stambaugh, David Krell, and Frank A, Capuzzi, David F. Krell (ed), New York, Harper & Row, 1984 ("The Eternal Recurrence of the Same and the Will to Power").

Nietzsche II, Pfullingen, Günther Neske, 1961 [1939–46], also GA 6.2, Frankfurt a/M, Vittorio Klostermann, 1997. Parts translated in *Nietzsche III: The Will to Power as Knowledge and Metaphysics*, trans. Joan Stambaugh, David F. Krell, and Frank A, Capuzzi, David F. Krell (ed), New York, Harper & Row, 1984 ("Nietzsche's Metaphysics"), *Nietzsche IV: Nihilism*, trans. Frank Capuzzi, David F. Krell (ed), New York, Harper and Row, 1982 ("European Nihilism" and "Nihilism as Determined by the History of Being"), and *The End of Philosophy*, trans. Joan Stambaugh, New York, Harper & Row, 1973 ("Metaphysics as History of Being," "Sketches for a History of Being as Metaphysics." and "Recollections in Metaphysics").

Introduction to Metaphysics, trans. Ralph Manheim, New Haven, Yale University Press, 1959; *Einführung in die Metaphysik*, Tübingen, Max Niemeyer, 1953, also GA Band 40, Vittorio Klostermann, Frankfurt a/M, 1983 [1935].

On Time and Being, trans. Joan Stambaugh, New York, Harper & Row, 1972; *Zur Sache des Denkens*, Tübingen, Max Niemeyer, 1969 [1962–64].

"Art and Space," trans. Charles H. Seibert, *Man and World* 3 (1973): 3-8; *Die Kunst und der Raum*, St. Gallon, Erker, 1969.

What Is Called Thinking?, trans. J. Glenn Gray, New York, Harper & Row, 1968; *Was heisst Denken?*, Tübingen, Max Niemeyer, 1971 [1951-2].

Selected Heidegger Lecture Courses Published Posthumously

Phänomenologie und transzendentale Wertphilosophie [*Phenomenology and Transcendental Value Philosophy*], in GA 56/57, Frankfurt a/M, Vittorio Klostermann, 1987 [1919].

Phänomenologie der Anschauung und des Ausdrucks. Theorie der philosophischen Begriffsbildung [*Phenomenology of Intuition and Expression. Theory of Philosophical Concept Formation*], GA 59, Frankfurt a/M, Vittorio Klostermann, 1993 [1920].

Phenomenological Interpretations of Aristotle: Initiation into Phenomenological Research, trans. Richard Rojcewicz, Bloomington, Indiana University Press, 2001; *Phänomenologische Interpretationen zu Aristoteles. Einführung in die phänomenologische Forschung*, GA 61, Frankfurt a/M, Vittorio Klostermann, 1985 [1921/22].

Ontology: The Hermeneutics of Facticity, trans. John van Buren, Bloomington, Indiana University Press, 1999; *Ontologie. Hermeneutik der Faktizität*, GA 63, Vittorio Klstermann, Frankfort a/M, 1988 [1923]

History of the Concept of Time. Prolegomena, trans. Theodore Kisiel, Bloomington, Indiana University Press, 1985; *Prolegomena zur Geschichte des Zeitbegriffs*, GA 20, second edition, Vittorio Klostermann, Frankfurt a/M, 1988 [1925].

Basic Problems of Phenomenology, trans. Albert Hofstadter, Bloomington, Indiana University Press, 1982; *Die Grundprobleme der Phänomenologie*, GA Band 24, second edition, Vittorio Klostermann, Frankfurt a/M, 1989 [1927].

Phenomenological Interpretation of Kant's Critique of Pure Reason, trans. Parvis Emad and Kenneth Maly, Bloomington, Indiana University Press, 1997; *Phänomenologische Interpretation von Kants Kritik der reinen Vernunft*, Frankfurt a/M, 1987 [1927/28].

Fundamental Concepts of Metaphysics: World, Finitude, Solitude, trans. William McNeill and Nicholas Walker, Bloomington, Indiana University Press, 1995; *Die Grundbegriffe der Metaphysik. Welt—Endlichkeit—Einsamkeit*, GA 29/30, Frankfurt a/M, Vittorio Klostermann, 1983 [1929/30].

On the Essence of Truth: On Plato's Parable of the Cave and the Theaetetus, trans. Ted Sadler, London, Continuum, 2002; *Vom Wesen der Wahrheit. Zu Platons Höhlengleichnis und Theätet*, GA 34, Frankfurt a/M, Vittorio Klostermann, 1988 [1931/32].

Hölderlins Hymnen "Germanien" und "Der Rhein," GA Band 39, Vittorio Klostermann, Frankfurt a/M, 1980 [1934/35].

What Is a Thing?, trans. W. B. Barton, Jr. and Vera Deutsch, Chicago, Henry Regnery Company, 1967; *Die Frage nach dem Ding. Zu Kants Lehre von den transzendentalen Grundsätzen*, GA 41, second edition, Vittorio Klostermann, Frankfurt a/M, 1984 [1935/36].

Schelling's Treatise on the Essence of Human Freedom, trans. Joan Stambaugh, Athens, Ohio University Press, 1984; *Schelling: Vom Wesen der menschlichen Freiheit*, GA Band 42, Vittorio Klostermann, Frankfurt a/M, 1988 [1936].

Hölderlins Hymne "Andenken", GA Band 52, Vittorio Klostermann, Frankfurt a/M, 1982 [1941/42].

Hölderlin's Hymn "The Ister," trans. William McNeill and Julia Davis, Bloomington, Indiana University Press, 1996; *Hölderlins Hymne "Der Ister"*, GA Band 53, Vittorio Klostermann, Frankfurt a/M, 1984 [1942].

Parmenides, trans. Andre Schuwer and Richard Rojcewicz, Bloomington, Indiana University Press, 1992; *Parmenides*, GA Band 54, Vittorio Klostermann, Frankfurt a/M, 1982 [1942/43].

Other Texts of Heidegger

Contributions to Philosophy (From Enowning), trans. Parvis Emad and Kenneth Maly, Bloomington, Indiana University Press, 1999; *Beiträge zur Philosophie. (Vom Ereignis)*, GA Band 65, Vittorio Klostermann, Frankfurt a/M, 1989 [1936–38].

"Only a God Can Save Us Now: An Interview with Martin Heidegger," trans. M.P. Alter and J.D. Caputo, in *Philosophy Today* 20, 1976: 267-84; "Nur ein Gott kann uns Retten," *Der Spiegel* 23 (May 1978).

"The Self-Assertion of the German University" and "The Rectorate 1933/34: Facts and Thoughts," *Review of Metaphysics* 38 (1985): 468-502; *Die Selbstbehauptung der deutschen Universität: Rede, gehalten bei der feierlichen Übernahme des Rektorats der Universität Freiburg i. Br. am 27.5.1933. Das Rektorat 1933/34: Tatsachen und Gedanken*, Vittorio Klostermann, Frankfurt a/M, 1983 [1933, 1945].

Schneeberger, Guido, *Nachlese zu Heidegger. Dokumente zu seinem Leben und Denken*, Buchdruckerei AG, Bern, 1962.

Selected General Secondary Works on Heidegger

Blattner, William, *Heidegger's Temporal Idealism*, New York, Cambridge University Press, 1999.

Caputo, John, *The Mystical Element in Heidegger's Thought*, New York, Fordham University Press, 1978.

Clark, Timothy, *Martin Heidegger*, London, Routledge, 2002.

Dallmayr, Fred, *The Other Heidegger*, Ithaca, Cornell University Press, 1993.

Dreyfus, Hubert, L., *Being-in-the-World: A Commentary on Heidegger's* Being and Time*, Division One*, Cambridge, MA, MIT Press, 1991.

_____, "On the Ordering of Things: Being and Power in Heidegger and Foucault," in *Michel Foucault: Philosopher*, Timothy J. Armstrong (ed), New York, Routledge, 1992, pp. 80-95.

Dreyfus, Hubert L. and Mark Wrathall (ed), *The Blackwell Companion to Heidegger*, Oxford, Blackwell, 2004.

Figel, Günter, *Martin Heidegger zur Einführung*, Hamburg, Junius Verlag, 2003.

Gethmann, Carl Friedrich, *Dasein, Erkennen und Handeln. Heidegger im phänomenologischen Kontext*, Berlin, Walter de Gruyter, 1993.

Guignon, Charles, *Heidegger and the Problem of Knowledge*, Indianapolis, Hackett, 1983.

Lacoue-Labarthe, Philippe, *Heidegger, Art and Politics*, trans. Chris Turner, Oxford, Blackwell, 1990; *La fiction du politique*, Paris, Christian Bourgois, 1988.

Olafson, Frederick A., *Heidegger and the Philosophy of Mind*, New Haven, Yale University Press, 1987.

Ott, Hugo, *Martin Heidegger: A Political Life*, trans. Allan Blunden, New York, Basic Books, 1993; *Martin Heidegger: Unterwegs zur seiner Biographie*, Frankfurt a/M, Campus-Verlag, 1988

Pöggler, Otto, *Martin Heidegger's Path of Thinking*, trans. Daniel Magurshak and Sigmund Barber, Atlantic Highlands, NJ, Humanities Press International, 1989; *Der Denkweg Martin Heideggers*, second edition, Pfullingen, Günther Neske, 1983.

Polt, Richard, *Heidegger*, London, UCL Press, 1999.

____, *The Emergency of Being*, Ithaca, Cornell University Press, 2006.

Rockmore, Tom and Joseph Margolis (ed), *The Heidegger Case: On Philosophy and Politics*, Philadelphia, Temple University Press, 1992.

Schürmann, Rainer, *Le Principe d'anarchie: Heidegger et la question de l'agir*, Paris, Editions de Seuil, 1982. Enlargened version published as *Heidegger on Being and Acting: From Principles to Anarchy*, Bloomington, Indiana University Press, 1987.

Thomson, Iain, *Heidegger on Ontotheology: Technology and the Politics of the University*, New York, Cambridge University Press, 2005.

Tugendhat, Ernst, *Der Wahrheitsbegriff bei Husserl und Heidegger*, Berlin, Walter de Gryuter, 1984.

Young, Julian, *Heidegger's Later Philosophy*, Cambridge, Cambridge University Press, 2002.

Wrathall, Mark, *How to Read Heidegger*, New York, W.W. Norton, 2006.

Selected Secondary Works on Heidegger's Ideas on Space

Arisaka, Yoko, "On Heidegger's Theory of Space: A Critique of Dreyfus," *Inquiry* 38, No. 4 (1995), pp. 455-67.

Casey, Edward, "Heidegger In and Out of Place," in *Heidegger: A Centenary Appraisal*, Pittsburgh, Silverman Phenomenology Center, 1990, pp. 62-98.

____, *The Fate of Place: A Philosophical History*, Berkeley, University of California Press, 1997.

Cloud-Cukoo-Land: International Journal of Architectural Theory 1998, No. 2, issue on "Bauen Wohnen Denken." Contains Gerd Achenbach, "Bauen

Wohnen Nachdenken," Joseph Burton, "Philosophische Unterschiede in den Gedanken von Louis I. Kahn and Martin Heidegger," Burkhard Diella, "Ein Denkweg an den anderen Anfang des Wohnens. Eine Interpretation von Heideggers Vortrag '*Bauen Wohnen Denken*,'" Gunter A. Dittmar, "Architektur als Wohnen und Bauen—Entwurf als ontologischer Akt," Hans Friesen, "Heideggers Architekturtheorie und die Moderne," Eduard Führ, "'genius loci'. Phänomen oder Phantom?," Karsten Harries, "Unterwegs zur Heimat," Dörte Kuhlmann, "Der Geist des (W)ortes," Alberto Pérez-Goméz, "Dwelling on Heidegger: Architecture as Mimetic *Techno-Poiesis*," David Seamon, "Heideggers Verständis von Wohnen konkretisiert: Die Beiträge von Thomas Thiis-Evensen and Christopher Alexander," Georg Christop Tholen, "Der Ort des Raums. Heideggers Kant-Lektüre und ihre Aktualität," and Svetozar Zavarihin, "Wohnen als Daseinsweise."

Dreyfus, Hubert, *Being-in-the-World: A Commentary on Heidegger's* Being and Time, *Division One*, Cambridge, MA, MIT Press, 1991, chapter seven.

Elden, Stuart, *Mapping the Present: Heidegger, Foucault and the Project of a Spatial History*, London, Continuum, 2001.

Fell, Joseph, *Heidegger and Sartre: An Essay on Being and Place*, New York, Columbia University Press, 1979.

Franck, D., *Heidegger et la Problème de l'Espace*, Paris, Editions de Minuit, 1986.

Harrison, Paul, "The space between us: Opening remarks on the concept of dwelling," *Environment and Planning D: Society and Space* (2007).

Malpas, Jeff, "Uncovering the Space of Disclosedness: Heidegger, Technology, and the Problem of Spatiality in Being and Time," in Jeff Malpas and Mark Wrathall (ed), *Heidegger, Modernity, and Authenticity: Essays in Honor of Hubert L. Dreyfus*, Volume 1, Cambridge, MA, MIT Press, 1999, pp. 205-28.

____, *Heidegger's Topology: Being, Place, World*, Cambridge, MA, MIT Press, 2006.

Müller, Christian, *Der Tod als Wandlungsmitte: Zur Frage nach Entscheidung, Tod und letztem Gott in Heideggers "Beiträge zur Philosophie"*, Berlin, Duncker & Humblot, 1999.

Neu, Daniela, *Die Notwendigkeit der Gründung im Zeitalter der Dekonstruktion: Zur Gründung in Heideggers "Beiträge zur Philosophie" unter Hinzuziehung der Derridaschen Dekonstruktion*, Berlin, Duncker & Humblot, 1997.

Neumann, Günther, *Die phänomenologische Frage nach dem Ursprung der mathematisch-naturwissenschaftlichen Raumauffassung bei Husserl und Heidegger*, Berling, Duncker & Humbolt, 1999.

Pickles, John, *Phenomenology, Science and Geography: Spatiality and the Human Sciences*, Cambridge, Cambridge University Press, 1985.

Pöggler, Otto, *Martin Heidegger's Path of Thinking*, trans. Daniel Magurshak and Sigmund Barber, Atlantic Highlands, NJ, Humanities Press International, 1989, chapter ten; *Der Denkweg Martin Heideggers*, second edition, Pfullingen, Günther Neske, 1983.

Polt, Richard, "Evoking the Momentous Site: Time-Space in the *Contributions to Philosophy*," paper presented at the 2003 North American Heidegger Conference, Old Dominion University, Norfolk, VA.

Steelwater, Eliza, "Mead and Heidegger: Exploring the Ethics and Theory of Space, Place, and Environment," in Andrew Light (ed), *Philosophy and Geography I: Space, Place, and Environmental Ethics*, Lanham, Rowman and Littlefield, 1997, pp. 189-207.

Vallega, Alejandro A., *Heidegger and the Issue of Space: Thinking on Exilic Grounds*, University Park, PA, The Pennsylvania State University Press, 2003.

Villel-Petit, Maria, "Space According to Heidegger—Some Guidelines," *Etudes Philosophiques* 2 (1981): 189-210.

_____, "Heidegger's Conception of Space," in C. Macann (ed), *Critical Heidegger*, London, Routledge, 1996, pp. 134-58.

Wohlfart, Günter, *Der Augenblick*, Freiburg, Alber, 1982.

Very Selected Texts from Heidegger's General Legacy

Betti, Emilio, *Zur Grundlegung einer allgemeinen Auslegungslehre*, Tübingen, J.C.B. Mohr, 1954.

Binswanger, Ludwig, *Grundformen und Erkenntnis menschlichen Daseins*, second edition, München, E. Reinhardt, 1962 [1942].

Borgmann, Albert, *Technology and the Character of Contemporary Life*, Chicago, University of Chicago Press, 1984.

Boss, Medard, *Psychoanalysis and Daseinanalysis*, trans. Ludwig B. Lefebre, New York, Basic Books, 1963; *Psychoanalyse und Daseinsanalytik*, Bern, Huber, 1957.

_____, *Existential Foundations of Medicine and Psychology*, trans. S. Conway and A. Cleaves, Northvale, Jason Aronson, 1979; *Grundriss der Medizin und der Psychologie*, Bern, Huber, 1971.

Bultmann, Rudolf, "The New Testament and Myth," in *Kerygma and myth; a theological debate*, trans. Reginald H. Fuller, London, S.P.C.K., 1953; "Neuer Testament u. Mythos," in *Kerygma und Mythos*, Hans Werner Bartsch (ed), Evangelischer Verlag, Hamburg, Volksdorf, 1948. This book contains responses to Bultmann's essay and counter-responses by Bultmann.

Carman, Taylor, *Heidegger's Analytic: Interpretation, Discourse and Authenticity in Being and Time*, New York, Cambridge University Press, 2003.

Davison, Aidan, *Technology and the Contested Meanings of Sustainability*, Albany, State University of New York Press, 2001.

Derrida, Jacques, *Of Grammatology*, trans. Gayatri Chakravorty Spivak, corrected edition, Baltimore, Johns Hopkins Unversity Press, 1993; *De la Grammatologie*, Paris Éditions de Minuit, 1967.

_____, *Margins of Philosophy*, trans. Alan Bass, second edition, Chicago, University of Chicago Press, 1985; *Marges de la Philosophie*, Paris, Éditions de Minuit, 1972.

Evernden, Nick, *The Natural Alien: Humankind and Environment*, Toronto, University of Toronto Press, 1985.

Gadamer, Hans-Georg, *Truth and Method*, trans. Joel Weinsheimer, and Donald G. Marshall, second, revised edition, New York, Crossroad, 1985; *Wahrheit und Methode*, fourth edition, Tübingen, J.C.B. Mohr, 1975 [1960].

Giddens, Anthony, *Central Problems in Social Theory*, Berkeley, University of California Press, 1979.

Gosden, Christopher, *Social Being and Time*, Oxford, Blackwell, 1994.

Grange, Joseph, "On the Way Toward Foundational Ecology," *Soundings* 60 (1977): 135-49.

Greene, Marjorie, *Approaches to a Philosophical Biology*, New York, Basic Books, 1968.

Haugeland, John, *Having Thought: Essays on the Metaphysics of Mind*, Cambridge, MA, Harvard University Press, 1998.

Hoy, David C., *The Critical Circle: Literature, History, and Philosophical Hermeneutics*, Berkeley, University of California Press, 1978.

Ingold, Tim, *The Perception of the Environment: Essays in Livelihood, Dwelling, and Skill*, London, Routledge, 2000.

Lipps, Hans, *Untersuchungen zu einer hermeneutischen Logik*, Frankfurt am Main, Klostermann, 1959.

Maglida, Robert, *Phenomenology and Literature*, West Lafayette, Purdue University Press, 1977.

Merleau-Ponty, Maurice, *Phenomenology of Perception,* trans. Colin Smith, London, Routledge and Kegan Paul, 1962; *Phénoménolgie de la Perception*, Paris, Éditions Gallimard, 1945.

Ott, Heinrich, *Denken und Sein: Der Weg Martin Heideggers und der Weg der Theologie*, Zollikon, Evangelischer Verlag, 1959.

Palmer, Richard, *Hermeneutics*, Evanston, Northwestern University Press, 1969.

Plessner, Helmuth, *Die Stufen des Organischen und der Mensch*, third edition, Berlin, Walter de Gruyter, 1975 [1928].

Rahner, Karl, *Spirit in the World*, trans. William Pych, New York, Herder & Herder, 1968; *Geist in Welt*, München, Kösel-Verlag, 1957.

Ricoeur, Paul, "The Task of Hermeneutics," *Philosophy Today* 17 (1973): 112-28; "La tâche de l'herméneutique," in *Exegesis: Problèmes de méthode et exercices de lecture*, François Bovon and Grégoire Rouiller (ed), Neuchâtel, Delachaux et Niestlé, 1975, pp. 179-200.

Rouse, Joseph, *Knowledge and Power: Toward a Political Philosophy of Science*, Ithaca, Cornell Unversity Press, 1987.

Taylor, Charles, "Heidegger, Language, and Ecology," in Hubert L. Dreyfus and Harrison Hall (ed), *Heidegger: A Critical Reader*, Oxford, Blackwell, 1992, pp. 247-69.

Thomas, Julian, *Time, Culture, and Identity: An Interpretive Archaeology*, London, Routledge, 1996.

Tillich, Paul, *The Courage to Be*, New Haven, Yale University Press, 1952.

Zimmerman, Michael, "Toward a Heideggerian *Ethos* for Radical Environmental-ism," *Environmental Ethics* 5 (1983), pp. 99-131.

Selected Texts from the Legacy of Heidegger's Ideas on Space

Arendt, Hannah, *The Human Condition*, Chicago, University of Chicago Press, 1958.

Bachelard, Gaston, *The poetics of space*, trans. Maria Jolas, Boston, Beacon Press, 1964; *La poétique de l'espace*, Paris, Presses Universitaires de France, 1958.

Bender, Barbara, *Stonehenge: Making Space*, Warwick, Berg, 1998.

Berque, Augustin, *Être humains sur la Terre. Principes d'éthique de l'écoumène*, Paris, Gallimard, 1996.

Binswanger, Ludwig, "Der Raum-Problem in der Psychopathologie," *Zeitschrift für Neurologie* (1933): 145ff.

Bollnow, Otto, *Mensch und Raum*, second edition, Stuttgart, Verlag W. Kohlhammer, 1971 [1963].

Bourdieu, Pierre, *Outline of a Theory of Practice*, trans. Richard Nice, Cambridge, Cambridge University Press, 1972; *Esquisse d'une théorie de la pratique, précédé de trois études d'ethnologie kabyle*, Genève, Libraire Droz, 1972.

____, *The Logic of Practice*, trans. Richard Nice, Stanford, Stanford University Press, 1990; *Le sens pratique*, Paris, Éditions de Minuit, 1980.

Buttimer, Anne, "Grasping the Dynamism of the Lifeworld," *Annuals of the Association of American Geographers* 66, No. 2 (1976): 277-92.

Casey, Edward, *Getting Back Into Place: Toward a Renewed Understanding of the Place-World*, Bloomington, Indiana University Press, 1993.

Cloke, Paul and Owain Jones, "Dwelling, Place and Landscape: an Orchard in Somerset," *Environment and Planning A* 33 (2001): 649-66.

Crick, Bernard, *In Defence of Politics*, Chicago, University of Chicago Press, 1962.

Derrida, Jacques, "Point de Folie—Maintenant L'Architecture," trans. Kate Linker, *AA Files* 12 (1986): 65-75.

____, *Of Hospitality: Anne Dufourmantelle Invites Jacques Derrida to Respond*, trans. Rachel Bowlby, Stanford, Stanford University Press, 2000; *De l'hospitalité: Anne Dufourmantelle invite Jacques Derrida à répondre*, Paris, Calmann-Lévy, 1997.

Dürckheim, Graf Karlfried von, "Untersuchungen zum Gelebten Raum," in Felix Krüger (ed), *Psychologische Optik*, München, C.H. Beck, 1932, pp. 383-480.

Elden, Stuart, *Mapping the Present: Heidegger, Foucault and the Project of a Spatial History*, London, Continuum, 2001.

Greene, Marjorie, "Landscape," in Ronald Bruzina and Bruce Wilshire (ed), *Phenomenology: Dialogues and Bridges*, Albany, SUNY Press, 1982, pp. 55-60.

Gölz, Walter, *Dasein und Raum*, Tübingen, Max Niemeyer, 1970.

Harries, Karsten, *The Ethical Function of Architecture*, Cambridge, MA, MIT Press, 1997.

Harvey, David, *Justice, Nature, and the Geography of Difference*, Oxford, Blackwell, 1996.

Ingold, Tim, *The Perception of the Environment: Essays in Livelihood, Dwelling, and Skill*, London, Routledge, 2000.

Irigaray, Luce, *The Forgetting of Air in Martin Heidegger*, trans. Mary Beth Mader, Austin, University of Texas Press, 1991; *L'oubli de l'air chez Martin Heidegger*, Paris, Editions de Minuit, 1983.

Kolb, David, *Postmodern Sophistications: Philosophy, Architecture, and Tradition*, Chicago, University of Chicago Press, 1990.

Kruse, Lenelis, *Räumliche Umwelt: Die Phänomenolgoie des räumlichen Verhaltens als Beitrag zu einer psychologischen Umwelttheorie*, Berlin, Walter de Gruyter, 1974.

Laclau, Ernesto and Chantal Mouffe, *Hegemony and Socialist Strategy: Toward a Radical Democratic Politics*, Cambridge, Verso, 1985.

Lassen, H, *Beiträge zu einer Phänomenologie und Psychologie der Anschauung*, 1939.

Lefebvre, Henri, *The Production of Space*, trans. Donald Nicholson-Smith, Oxford, Blackwell, 1991; *La production de l'espace*, Paris, Anthropos, 1974.

Lévinas, Emmanuel, *Totality and Infinity: an essay on exteriority*, trans. Alphonso Lingis, Pittsburgh, Duquesne University Pres, 1969; *Totalité et infini. Essai sur l'extériorité*, La Haye, Nijoff, 1968.

Lévy, Jacques and Michel Lussault (ed), *Logiques de l'espace, espirit des lieux. Géographies à Cerisy*, Paris, Belin, 2000.

Lussault, Michel, "Action(s)!," in *Logiques de l'espace, espirit des lieux. Géographies à Cerisy*, Jacques Lévy and Michel Lussault (ed), Paris, Belin, 2000, pp. 11-36.

Macnaghton, Phil and John Urry, *Contested Natures*, London, Sage, 1998.

Malpas, Jeff, *Place and Experience*, Cambridge, Cambridge University Press, 1999.

Merleau-Ponty, Maurice, *Phenomenology of Perception*, trans. Colin Smith, London, Routledge and Kegan Paul, 1962, part one, chapter three and part two, chapter two; *Phénoménolgie de la Perception*, Paris, Éditions Gallimard, 1945.

Meiss, Pierre van, *Elements of Architecture: From Form to Place*, London, E & FN Span, 1991; *De la Forme au Lieu*, Lausanne, Presses Polytechniques romandes, 1986.

Minkowski, Eugène, *Lived Time: Phenomenological and Psychopathological Studies*, trans. Nancy Metzel, Evanston, Northwestern University Press, 1970; *Le temps vécu. Etudes phénoménologiques et psychopathologiques*, Paris, Presses Universitaires de France, 1995 [1933].

Mugerauer, Robert, *Interpretations on Behalf of Place: Environmental Displacements and Alternative Responses*, Albany, SUNY Press, 1995.

_____, *Interpreting Environments: Tradition, Deconstruction, Hermeneutics*, Austin, University of Texas Press, 1995.

Nancy, Jean-Luc, *Being Singular Plural*, trans. Robert D. Richardson and Anne E. O'Byrne, Stanford, Stanford University Press, 2000; *Être singulier pluriel*, Paris, Éditions Galilée, 1996.

Norberg-Schulz, Christian, *Existence, Space, and Architecture*, New York, Praeger Publishers, 1971.

_____, "Kahn, Heidegger and the Language of Architecture," *Oppositions* 18 (1979): 28-47.

_____, *Genius Loci: Towards a Phenomenology of Architecture*, New York, Rizzoli, 1980. (First published in Italian as *Genius Loci—paesaggio, ambiente, architettura*, Milano, Gruppo Editoriale Electa, 1979).

_____, *The Concept of Dwelling: On the way to figurative architecture*, New York, Electa/Rizzoli, 1985.

Otto, Maria A. C., *Der Ort. Phänomenolgische Variationen*, Freiburg, Karl Alber, 1992.

Relph, Edward, *Place and Placelessness*, London, Pion, 1976.

_____, *Rational Landscapes and Humanistic Geography*, London, Croom Helm, 1981.

Rouse, Joseph, *How Scientific Practices Matter: Reclaiming Philosophical Naturalism*, Chicago, University of Chicago Press, 2003.

Schatzki, Theodore, "Spatial Ontology and Explanation," *Annals of the Association of American Geographers* 81, No. 4 (1993): 650-70.

_____, *The Site of the Social: A Philosophical Exploration of the Constitution of Social Life and Change*, University Park, The Pennsylvania University State Press, 2002.

Seamon, David, *A Geography of the Lifeworld: Movement, Rest and Encounter*, New York, St. Martin's Press, 1979.

_____, (ed), *Dwelling, Seeing, and Designing: Toward a Phenomenological Ecology*, Albany, State University of New York Press, 1993.

Seamon, David and Robert Mugerauer, *Dwelling, Place & Environment: Towards a Phenomenology of Person & World*, New York, Columbia University Press, 1985.

Simonsen, Kirsten, "Space, culture and economy - A question of practice," *Geografiska Annaler*, 83B, No. 1 (2000): 41-53.

Soja, Edward, *Postmodern Geographies: The Reassertion of Space in Critical Social Theory*, London, Verso, 1989.

Spinosa, Charles, Fernando Flores, and Hubert L. Dreyfus, *Disclosing New Worlds: Entrepreneurship, Democratic Action, and the Cultivation of Solidarity*, Cambridge, MA, MIT Press, 1997.

Strohmayer, Ulf, "The event of space: geographic allusions in the phenomenological tradition," *Environment and Planning D: Society and Space* 16 (1998): 105-21.

Ströker, Elisabeth, *Philosophische Untersuchungen zum Raum*, Frankfurt a/M, Vittorio Klostermann, 1965.

Strauss, Erwin, "Die Formen des räumlichen Erlebens: Ihre Bedeutung für die Motorik und die Wahrnehmung," *Der Nervenarzt* 3, No. 11 (1930).

Taylor, Charles, "Interpretation and the Sciences of Man," in his *Philosophy and the Human Sciences: Philosophical Papers 2*, Cambridge, Cambridge University Press, 1985 [1971], pp. 15-58.

Thomas, Julian, *Time, Culture, and Identity: An Interpretive Archaeology*, London, Routledge, 1996.

Thrift, Nigel, "Steps Toward an Ecology of Place," in Doris Massey, J. Allen, and P. Sarre (ed), *Human Geography Today*, Cambridge, Polity, 1999, pp. 295-322.

Tilley, Christopher, *A Phenomenology of Landscape. Places, Paths and Monuments*, Oxford, Berg, 1994.

Waldenfels, Bernhard, *In den Netzen der Lebenswelt*, Frankfurt a/M, Suhrkamp, 1985.

Young, Julian, "What is Dwelling? The Homelessness of Modernity and the Worlding of the World," in *Heidegger, Authenticity, and Modernity: Essays in Honor of Hubert L. Dreyfus*, Volume I, Mark Wrathall and Jeff Malpas (ed), Cambridge, MA, MIT Press, pp. 187-203.

Miscellaneous

Braig, Carl, *Vom Sein. Abriss der Ontologie*, Freiburg, Herder, 1896.

Brentano, Franz, *On the Several Senses of Being in Aristotle*, trans. Rolf George, Berkeley, University of California Press, 1975; *Von der mannigfachen Bedeutung des Seienden nach Aristoteles*, Freiburg, Herder, 1862.

Husserl Edmund, *Logical Investigations*, Volumes One and Two, trans. J.N. Findlay, London, Routledge and Kegan Paul, 1970; *Logische Untersuchungen. Erste Teil: Prolegomena zur reinen Logik*, Tübingen, Max Niemeyer, 1968 [1900], *Logische Untersuchungen. Zweite Teil: Untersuchungen zur Phänomenologie und Theorie der Erkenntnis*, Tübingen, Max Niemeyer, 1968 [1901].

Thomas, Julian, "Archaeologies of Place and Landscape," in Ian Hodder (ed), *Archaeological Theory Today*, Cambridge, Polity, 2001, pp. 165-186.

Whorf, Benjamin, *Language, Thought, and Reality: Selected Writings of Bejnamin E. Whorf*, John B. Carroll (ed), Cambridge, MA, MIT Press, 1967.

Williams, Raymond, *The Long Revolution*, London, Chatto & Windus, 1961.

Appendix: German Quotations

Chapter Three. Basic Philosophical Ideas

a   *"Das 'Wesen' des Daseins liegt in seiner Existenz."* (SZ 42)

b   "Das Seiende, das wesenhaft durch das In-der-Welt-sein konstituiert wird, *ist* selbst je sein 'Da'." (SZ 132)

c   "Der Mensch ist der Hirt des Seins." ("Der Brief über Humanismus," in *Wegmarken*, s. 328)

d   "...ist die Sprache das Haus des Seins." ("Der Brief über Humanismus," in *Wegmarken*, s. 330)

e   "Das Wesentliche ist, dass wir mitten in der Vollendung des Nihilismus stehen, dass Gott 'todt' ist und jeder Zeit-Raum für die Gottheit verschüttet. Dass sich gleichwohl die Verwindung des Nihilismus ankündigt im dichtenden Denken und Singen des Deutschen..." (*Die Selbstbehauptung der deutschen Universität: Rede, gehalten bei der feierlichen Übernahme des Rektorats der Universität Freiburg i. Br. am 27.5.1933. Das Rektorat 1933/34: Tatsachen und Gedanken*, s. 39)

f   "...indem Hölderlin das Wesen der Dichtung neu stiftet, bestimmt er erst eine neue Zeit. Es ist die Zeit der entflohenen Götter *und* des kommenden Gottes." (*Erläuterungen zu Hölderlins Dichtung*, s. 47)

Chapter Four. Space, Spatiality, and Society

a   "...das, *worin* ein faktisches Dasein als dieses 'lebt'." (SZ 65)

b   "Die phänomenale Struktur der Weltlichkeit des Raumes bezeichnen wir als das *Umhafte* an der Welt als *Umwelt*." (*Prolegomena zur Geschichte des Zeitbegriffs*, s. 308).

c   "…sofern er weltlich da ist, überhaupt von mir entfernt ist, überhaupt zu mir eine mögliche Nähe und Ferne hat…" (*Prolegomena zur Geschichte des Zeitbegriffs*, s. 309)

d   "Die jeweilige Hingehörigkeit entspricht dem Equipmentcharakter des Zuhandenen…" (SZ 102)

e   "…Wo des Wohin eines Hingehörens, Hingehens, Hinbringens, Hinsehens und dergleichen." (*Prolegomena zur Geschichte des Zeitbegriffs*, s. 314)

f   "So etwas wie Gegend muss zuvor entdeckt sein, soll das Anweisen und Vorfinden von Plätzen einer umsichtig verfügbaren Zeugganzheit möglich werden." (SZ 103)

g   "…hat…durch die weltmässige Bewandtnisganzheit…ihre eigene Einheit…Die 'Umwelt' richtet sich nicht in einem zuvorgegebenen Raum ein, sondern ihre spezifische Weltlichkeit artikuliert in ihrer Bedeutsamkeit den bewandtnishaften Zusammenhang einer jeweiligen Ganzheit von umsichtig angewiesenen Plätzen. Die jeweilige Welt entdeckt je die Räumlichkeit des ihr zugehörigen Raumes." (SZ 104)

h   "Diese gegendhafte Orientierung der Platzmannigfaltigkeit des Zuhandenen macht das Umhafte, das Um-uns-herum des umweltlich nächstbegegnenden Seienden aus." (SZ 103)

i   "…innan-, wohnen, habitare, sich aufhalten; 'am' bedeutet: ich bin gewohnt, vertraut mich, ich pflege etwas… 'ich bin' besagt wiederum: ich wohne, halte mich auf bei…der Welt, als dem so und so Vertrauten." (SZ 54)

j   "Weil das Dasein wesenhaft räumlich ist in der Weise der Ent-fernung, hält sich der Umgang immer in einer von ihm je in einem gewissen Spielraum entfernten 'Umwelt.'" (SZ 107)

k   "…Wobei eines ent-fernenden Seins bei…in eins mit dieser Ent-fernung." (SZ 107)

l   "Die Plätze und die umsichtig orientierte Platzganzheit des zuhandenen Equipments sinken zu einer Stellenmannigfaltigkeit für beliebige Dinge zusammen. Die Räumlichkeit des innerweltlich Zuhandenen verliert mit diesem ihren Bewandtnischarakter. Die Welt geht des spezifisch Umhaften verlustig, die Umwelt wird zur Naturwelt. Die 'Welt' als zuhandenes Equipmentganzes wird verräumlicht zu einem Zusammenhang von nur noch vorhandenen ausgedehnten Dingen." (SZ 112)

[m] "Die Kunst ist das Sich-ins-Werk-Setzen der Wahrheit." ("Der Ursprung des Kunstwerkes," in *Holzwege*, s. 24)

[n] "...Bahnen und Bezüge..., in denen Geburt und Tod, Unheil und Segen, Sieg und Schmach, Ausharren und Verfall—dem Menschenwesen die Gestalt seines Geschickes gewinnen." ("Der Ursprung des Kunstwerkes," in *Holzwege*, s. 27)

[o] "...Weile und Eile, ihre Ferne und Nähe, ihre Weite und Enge." ("Der Ursprung des Kunstwerkes," in *Holzwege*, s. 30).

[p] "Aufstellend eine Welt und herstellend die Erde ist das Werk die Bestreitung jenes Streites, in dem die Unverborgenheit des Seienden im Ganzen, die Wahrheit, erstritten ist." ("Der Ursprung des Kunstwerkes," in *Holzwege*, s. 41)

[q] "...ihren Stand und ihre Ständigkeit nimmt." ("Der Ursprung des Kunstwerkes," in *Holzwege*, s. 47)

[r] "Die Einrichtung der Wahrheit ins Werk ist das Hervorbringen eines...Seienden. ... Die Hervorbringen stellt dieses Seiende dergestalt ins Offene, dass das zu Bringende erst die Offenheit des Offenen lichtet, in das es hervorkommt." ("Der Ursprung des Kunstwerkes," in *Holzwege*, s. 48).

[s] "Das Wort durchmisst als der sinnliche Sinn die Weite des Spielraumes zwischen Erde und Himmel. Die Sprache hält den Bereich offen, in dem der Mensch auf der Erde unter dem Himmel das Haus der Welt bewohnt." (*Hebel, der Hausfreund*, s. 38)

[t] "...die Götter und der Tempel, der Fest und die Spiele, die Herrscher und der Rat der Alten, die Volksversammlung und die Streitmacht, die Schiffe und die Feldherrn, die Dichter und die Denker." (*Hölderlins Hymne "Der Ister"*, s. 101)

[u] "...die in sich gesammelte Stätte der Unverborgenheit des Seiendes im Ganzen" (*Parmenides*, s. 133)

[v] "...alles Seiende und alles Verhalten zum Seienden sich sammelt." (*Hölderlins Hymne "Der Ister"*, s. 106)

[w] "...die Stätte, das Da, worin und als welches das Da-sein als geschichtliches ist. Die πόλις ist die Geschichtsstätte, das Da, *in* dem, *aus* dem und *für* das die Geschichte geschieht." (*Einführung in die Metaphysik*, s. 161)

^x   "Die πόλις ist die Wesensstätte des geschichtlichen Menschen, das Wo, wohin der Mensch als ζῷον λόγον ἔχον gehört, das Wo, von woher allein ihm zugefügt wird der Fug, in den er gefügt ist." (*Parmenides*, s. 141)

^y   "…erste Lichtung des Offenen als der 'Leere.'" (*Beiträge zur Philosophie. (Vom Ereignis)*, s. 380)

^z   "Die Zeit-Raum is der berückend-entrückende sammelnde Umhalt…dessen Wesung in der Gründung des 'Da' durch das Da-*sein* (seine wesentliche Bahnen der Bergung der Wahrheit) geschichtlich wird." (*Beiträge zur Philosophie. (Vom Ereignis)*, s. 386).

^{aa}   "…die in sich zeitigend-räumend-gegenschwingende Augenblicksstätte des 'Zwischen,' als welches das Da-sein gegründet sein muss." (*Beiträge zur Philosophie. (Vom Ereignis)*, s. 387)

Chapter Five. Legacies

^a   "Nur wenn wir das Wohnen vermögen, können wir bauen." ("Bauen Wohnen Denken," in *Vorträge und Aufsätze*, s. 155)

Author Index

SOZIALGEOGRAPHISCHE BIBLIOTHEK

Herausgegeben von Benno Werlen

1. Christian Schmid:
   Stadt, Raum und Gesellschaft
   Henri Lefebvre und die Theorie der
   Produktion des Raumes. 2005. 344 S.,
   kt.   ISBN 978-3-515-08451-2

2. Roland Lippuner:
   Raum – Systeme – Praktiken
   Zum Verhältnis von Alltag, Wissen-
   schaft und Geographie. 2005. 230 S.,
   kt.   ISBN 978-3-515-08452-9

3. Ulrich Ermann:
   Regionalprodukte
   Vernetzungen und Grenzziehungen
   bei der Regionalisierung von Nah-
   rungsmitteln. 2005. 320 S. m. 15 Abb.,
   kt.   ISBN 978-3-515-08699-8

4. Antje Schlottmann:
   RaumSprache
   Ost-West-Differenzen in der Bericht-
   erstattung zur deutschen Einheit.
   Eine sozialgeographische Theorie.
   2005. 343 S., kt.
   ISBN 978-3-515-08700-1

5. Edgar Wunder:
   **Religion in der postkonfessionellen
   Gesellschaft**
   Ein Beitrag zur sozialwissenschaft-
   lichen Theorieentwicklung in der Re-
   ligionsgeographie. 2005. 366 S., kt.
   ISBN 978-3-515-08772-8

6. Theodore R. Schatzki:
   Martin Heidegger: Theorist of Space
   2007. 129 S., kt.
   ISBN 978-3-515-08956-2

FRANZ STEINER VERLAG STUTTGART